The Good Life of Helen K. Nearing ❦

The Good Life of
Helen K. Nearing

Margaret O. Killinger

UNIVERSITY OF VERMONT PRESS
Burlington, Vermont
Published by University Press of New England
Hanover and London

UNIVERSITY OF VERMONT PRESS
Published by University Press of New England,
One Court Street, Lebanon, NH 03766
www.upne.com

© 2007 by University Press of New England
Printed in the United States of America
5 4 3 2 1

Library of Congress Cataloging-in-Publication Data
Killinger, Margaret O.
The good life of Helen K. Nearing / Margaret O. Killinger.
   p.   cm.
Includes bibliographical references and index.
ISBN-13: 978-1-58465-628-9 (cloth : alk. paper)
ISBN-10: 1-58465-628-X (cloth : alk. paper)
1. Nearing, Helen.   2. Country life—Vermont.   3. Country life—Maine—
Harborside.   I. Title.
S521.5.V5K55 2007
630.92—dc22
[B]   2006102248

University Press of New England is a member of the Green Press Initiative. The paper used in this book meets their minimum requirement for recycled paper.

Frontispiece: Portrait of Helen Knothe Nearing, 1984. © photo by Lynn Karlin.

The author gratefully acknowledges permission to use the following material:

Interviews
Nancy Berkowitz, interviews with Margaret Killinger, tape recordings, March 3, 1999, and April 2, 2001.
Nancy Caudle-Johnson, interview with Margaret Killinger, tape recording, January 24, 2004.

(Continued on page 141)

To Eric and Meg

Contents ❧

Acknowledgments ❧

Jerry Goldman, a sculptor and former neighbor of the Nearings in Vermont, was one of the many remarkable people whom I met over the course of this project. Like the Nearings, Jerry built with Vermont stone, though he swirled his mortar in a Gaudí-esque fashion around the contour of each rock in such a way that he sculpted, more than built, his home. He fashioned a small chapel on his property in this style, decorating the curved interior walls with placards listing the names of his family members, close friends, and neighbors. Jerry provided natural light through fifteen stained-glass windows, each of which subtly revealed a letter, spelling around the circumference of the chapel: WE NEED EACH OTHER.

Recognizing at the end of this book the profound wisdom in Jerry's sentiment, I would like to thank many of the people whom I have needed and relied upon to help me complete my work. I would like to express my gratitude to professors at the University of Maine: Nathan Godfried, Richard Judd, Kristin Langellier, Martha McNamara, James Moreira, Charles Slavin, and especially Marli Weiner.

I would like to thank archivists Jeffrey Cramer of the Thoreau Institute and Sean Noel of the Howard Gottlieb Research Center of Boston University for their patience and able assistance with my research. And I would like to express my heartfelt appreciation to the Lindberg family, to Clifford and Dorrie Bean, and to David and Merle Trager for their hospitality and their enthusiasm for this project as they graciously welcomed me into their homes during numerous research trips.

I also would like to thank Nearing friends and family Nancy Berkowitz, Jeanne Gaudette, Nancy Caudle-Johnson, Rebecca and

x   Eugene Lepkoff, Robert Nearing, Greg Joly, Cornelia Tuttle, Lynn Karlin, and Richard Garrett. I regret that I cannot offer one last thanks to the late Gail Disney and the late Jerry Goldman, both of whom passed away just a few winters ago.

I would like to express sincere gratitude to my Saturday morning writers' group: Martha van der Voort, Cheryl Daigle, Ellen Mallory, Betty Baldwin, Tanya Baker, and Suzy Kaback. And I extend a special thanks to my editor at University Press of New England, Phyllis Deutsch, for her encouragement and her steady, skillful work.

Thank you to my family, the whole wide web, and especially my niece and research assistant Sarah Pandiscio, my parents Pete Killinger and the late Peggy Killinger, as well as my sister Leslie Killinger Yzaguirre and her boys Mario, Max, and Will. Finally, I offer my deepest thanks always to Eric and Meg Pandiscio—my good life and my home.

The Good Life of Helen K. Nearing

Prologue

ALONE IN HER STONE HOUSE on the coast of Maine, Helen Nearing stepped onto her second-floor balcony to sun herself. A wind gust suddenly blew her balcony door shut, locking her out, which surprised her, as locks were rarely engaged at Forest Farm. She remained on the balcony for several hours, hoping that a visitor might stop by, but none came. Remarkably strong, adventurous, and impetuous for a woman in her eighties, Nearing jumped from the balcony. She was severely scraped and bruised by the gravel below but did not break any bones. Helen Nearing thoroughly enjoyed telling and retelling the story of that jump, which was reminiscent of the many bold leaps that she had taken over the course of her colorful life.

Helen Knothe (pronounced "Ken-*oh*-thee") Nearing (1904–1995; hereafter referred to as HK before her marriage and HKN after) and her long-term homesteading partner Scott Nearing (1883–1983) gained iconic status in the 1970s as the grandparents of the late twentieth century back-to-the-land movement.[1] The Nearings remythified and popularized voluntary, self-sufficient living on the land through autobiographies, lectures, and the daily enactment of their homesteading life on their Forest Farm. They heralded organic gardening, vegetarianism, pacifism, and hard work as primary components of their "good life"—a term rooted in Aristotelian thought and appealing to the ongoing human search for an ideal existence. Sociologist Jeffrey Jacob claimed, "Perhaps no other modern homesteaders better exemplify the single-minded devotion to the principles of simple living in the country than Helen and Scott Nearing." Jacob believed that through back-to-the-land discipline, the Nearings found "liberation from the mass-consumption marketplace" and fulfillment through their "relationship to the land."[2]

The Nearings' seminal book, *Living the Good Life* (1954) offered a detailed account of their homesteading life in Vermont from 1932 to 1952. It provided a handbook to aspiring homesteaders, as the Nearings described in detail their daily work, as well as their dedication to personal liberation through rural living, their belief in nonviolence, and their deep affection for vegetables.[3] According to Jacob, their book read like their own "declaration of independence from corporate America." They proposed an ethic of simple living based upon production rather than consumption, allowing nature rather than the market economy to set the terms for their choices.[4] *Living the Good Life* was republished in 1970 and has been reprinted over thirty times and sold more than three hundred thousand copies.[5]

The Nearings wrote several other autobiographies, which might suggest that this biography of Helen Nearing is an unnecessary exercise given the extent to which her story already has been told. However, as is often the case with personal narratives, the Nearings' various renditions of their good life were replete with glaring omissions, inconsistencies, and embellishments. Autobiography is by definition a dubious source for history. Shari Benstock, in her essay "Authorizing the Autobiographical," argued that the writer of autobiography uses language as a symbolic system to construct the autobiographical subject, building a "psychic wall" or a "linguistic fortress between the writer and [her] interested readers."[6] Consistent with Benstock's argument, the Nearings' autobiographical texts were narratives that illustrated far-too-neat versions of their past personas, building a "linguistic fortress" or "psychic wall" that sealed off truths and obscured facts.

The autobiographical divide separating HKN from her audience was especially pronounced as she created and mythologized their good life story. For example, in her audiotape, *The Good Life of Helen Nearing* (1994), HKN made spurious claims regarding the construction of their homestead, inflating the Nearings' building contributions and denying the abundant help that they received.[7] Her most notable autobiographical omission occurred in her later memoir, *Loving and Leaving the Good Life* (1992), in which she described Scott Nearing's "simple passing and beautiful death."[8] She introduced the book with the disclaimer:

> This is not a work of fiction. Names, characters, places, and incidents are not a product of the author's imagination. Any resemblance to actual events, locales, or to persons, living or dead, is intended and perfectly true to the author's memory of them.[9]

However, she then offered a romantic, idealized version of his death that left out critical, personal details and completely excluded Nancy Berkowitz, a principal caregiver to Nearing throughout his dying process.

Such unrealistic portrayals in HKN's good life stories resonate with the Personal Narratives Group's description of autobiography in *Interpreting Women's Lives:*

> When talking about their lives, people lie sometimes, forget a lot, exaggerate, become confused, and get things wrong. Yet they *are* revealing truths. These truths don't reveal the past "as it actually was," aspiring to a standard of objectivity. They give us instead the truths of our experiences . . . We come to understand them only through interpretation, paying careful attention to the contexts that shape their creation and to the world views that inform them.[10]

Thus, in order for a critical biographer to rely upon HKN's personal narratives as historical resources, she or he must discern the truths

4

within misinformation and obscured facts, while also considering context and interpreting dubious claims.

Other resources beyond personal narratives become essential for an historical rendering of Helen Nearing's good life. Oral histories conducted with her family and friends, and photographs, diaries, and letters help to establish her historical context and add dimension and character to HKN as a biographical subject. Unfortunately for the researcher, after Scott Nearing's house was raided by government agents in Toledo, Ohio, during World War I, he destroyed all of his personal correspondences from that time forward. Therefore, only a handful of Helen's letters to Scott—those that he happened to include in his responses to her letters—remain available for study.[11] However, shortly before her death, HKN entrusted an extensive collection of her private papers to the Trust for Public Land. In 1998, the papers were lodged in the newly established Thoreau Institute at Walden Woods in Lincoln, Massachusetts, and made readily available to the public.

Sources such as the eulogistic, spiritual biographies of Helen Nearing written by her close friend Ellen LaConte also offer insight into HKN's life. The scholarly, critical work of Rebecca Kneale Gould on the art of modern homesteading and the Nearings' central role in it provides a vital, interpretive gloss. Yet the most essential archive for a biography of HKN remains the Nearings' Forest Farms. HKN's deep commitment to the natural world—she claimed that she was "born enamored of the earth"—and the Nearings' good life values of "simplicity, serenity, utility, and harmony" were exemplified by, and played out upon, their homesteads.[12]

Furthermore, the Forest Farm landscapes, particularly their final Forest Farm, provide a critical means for separating Helen Nearing's story from Scott Nearing's. One of the foremost challenges for a biographer of HKN is to highlight her unique contributions to the Nearings' good life story. HKN persistently resisted such recognition, maintaining a puzzling subservience and deference to her powerful partner. This biography of Helen Nearing charts her development as a free-standing individual and her creation of a primary role for herself within their work. With her deep spirituality, her keen sense of purpose, and her formidable marketing and promotional skills, HKN would move to the forefront of their good life, particularly in the context of Forest Farm and her daily work there.

Helen Nearing also separated herself from Scott (through her creative, tenacious telling and retelling of their story) as she established their remarkable legacy. In 2004, on the fiftieth anniversary year of its publication, Paul Greenberg of the *Boston Globe* wrote of *Living the Good Life* that it is "a work that never seems to leave us."[13] Editor of *Organic Gardening Magazine* Scott Mcyer claimed that the Nearings planted the "seed" for the organic food movement, the fastest-growing sector in the American food industry today.[14] MaryJane Butters, described as a present-day "farmer in the same way Martha Stewart is a housewife," remains indebted to the Nearings; she runs the million-dollar MaryJanesFarm organic food and goods mail-order company out of Idaho. Thousands of visitors attend her monthly Market Days at her farm as she capitalizes on a burgeoning "agritourism," originating with the Nearings' hospitality on their first homestead seventy years ago.[15]

This biography of Helen Knothe Nearing traces the evolution of this influential spokesperson for organic farming and for simple, purposeful living, particularly during the second half of the last century. It dispels certain myths but upholds the basic integrity of the good life that HKN created, promoted, and lived. As Helen Nearing expressed endless enthusiasm for the way of life she had chosen, this biographer believes that her good life story merits yet another retelling.[16]

CHAPTER 1

Spiritual Formation

The Theosophical Society
and Travels with Krishnamurti ✤

HELEN KNOTHE described hers as "an admirable family" to have chosen to be born into on the morning of February 23, 1904.[1] She was the second child of Frank and Maria Knothe, upper-middle-class, turn-of-the-century urbanites living at 212 West Fourteenth Street in New York City. Helen's older brother, Alex, had been born three years earlier, and her younger sister, Alice, would be born three years later. Frank Knothe worked as vice president of Knothe Brothers Company, a successful haberdashery business in New York City specializing in men's suspenders, belts, and pajamas. Stoic and authoritative, Frank Knothe was later described by his granddaughter as having a Germanic lack of warmth.[2]

Maria Knothe was a reserved woman of Dutch heritage, a former artist who had set aside her painting in order to raise her children. Helen Knothe identified closely with her mother's Dutch lineage. She learned to read and speak Dutch, lived in Amsterdam, wore Dutch clogs, and maintained ties to Dutch friends and family throughout her life. HK's attachment to Holland stemmed in part from her deep fondness for her mother. Though HK later proved to be a rebellious family member, she wrote that the one thing she and her siblings agreed upon was a great affection for their mother.[3] In a letter to Scott, Helen described being "one with the atmosphere in Holland." She claimed a particularly keen spiritual connection to the country and its people, as she believed that she was the reincarnation of the founder of Dutch theosophy, Petronella Catharina Meuleman van Ginkel (1841–1902), known as Piet Meuleman.[4]

In 1907, the Knothes changed primary residences, though they retained their apartment in New York, and moved to their "Jersey haven" at 74 Cottage Place in Ridgewood, New Jersey. Their house was an East Indian bungalow-style structure with broad verandahs, timbered ceilings, and cozy nooks for reading. The wooded Cottage Place property included extensive lawns, a tennis court, and a large organic garden to support the family's vegetarian diet.[5] Inspired by her parents' example, HK became a lifelong, ardent vegetarian.

In New Jersey, Maria Knothe was involved in the Ridgewood Women's League, the Society for the Prevention of Cruelty to Animals, and the women's organization of the town's Unitarian Church. Frank Knothe sang in a men's community chorus, served on the Ridgewood board of education, and organized speakers for political and intellectual debate at the local Arts Club and their Unitarian Church. The Knothe children thus were exposed early on to left-leaning, civic engagement.[6]

The children also were introduced to a nontraditional spirituality that diverged from their more mainstream Unitarianism. Frank and Maria Knothe were extremely active in the Theosophical Society, an offshoot of nineteenth-century Spiritualism inspired by the clairvoyant Russian mystic, Madame Blavatsky. Theosophy was a spiritual science derived from elements of occultism and various Eastern religions, as well as from modern science.[7] Belief in reincarnation, intellectual inquiry, meditation, prayer, and vegetarianism were central to theosophy, along with clairvoyant connection to a mystical brotherhood of Masters.

Theosophy's Eastern orientation, particularly Blavatsky's alleged link to Tibet and its religious symbolism, resonated with broader nineteenth-century trends. Interest in Asian religious studies had proliferated in Europe beginning in the late eighteenth century with the formation of the Royal Asiatic Society in London and with the nineteenth-century translations of Hindu scriptures and major Buddhist texts into French, German, and English. Blavatsky's contact with the "Masters" derived in part from the contemporary, popular séance traditions rooted in American Spiritualism and in the esoteric writings of British novelist Edward Bulwer Lytton (1803–1873). Historian Peter Washington contended: "It would not be unjust to say that her new religion was virtually manufactured from [Lytton's] pages."[8]

Despite Blavatsky's reliance upon Eastern mysticism and Western esotericism, she proved a unique religious character as she generated an exhaustive list of personal spiritual Masters. Master Morya was her principal contact with her Great White Brotherhood of Masters or Mahatmas—absolutely pure initiates who had undergone rigorous training in esoteric thought and who were endowed with supernatural powers as they inhabited both material and semi-material bodies. Her Masters directed the course of Blavatsky's travels, and in 1873, twenty-two years after Master Morya first made contact with her, these adepts told Blavatsky to move to New York City.[9]

At that time, Blavatsky's physical appearance would have been as eccentric as her religious persuasion. She was quite stout (eventually weighing over two hundred and forty pounds), with unruly hair and notably bulging, blue eyes. She wore "shabby fantastic clothes," rings on every finger, and an animal-head pouch around her neck for the cigarettes that she constantly smoked. Blavatsky tended to speak incessantly and sometimes crudely, and was unpretentious, capricious, vulgar, and humorous.[10] Ironically, her cofounder of the Theosophical Society, Henry Olcott, appeared as ordinary as Blavatsky was unusual.

Olcott was born in 1832 into a bourgeois New Jersey family. In the wake of a failed marriage and after a succession of jobs as an Ohio farmer, a Union Army colonel, and an attorney in New York City, Olcott became interested in Spiritualism and began attending séances at Mary Baker Eddy's farm, Chittenden, in Vermont. He wrote a series of articles for a New York–based Spiritualist paper, the *Daily Graphic*, about his experiences at Chittenden that interested Blavatsky. In turn, Blavatsky sought out Olcott at Chittenden in the summer of 1875 and proceeded to monopolize séances there, flaunting her powers and summoning "her uncle, two Russian servants, a Persian merchant, and a Kurdish warrior."[11]

With such displays, Blavatsky shattered the boundaries of Spiritualism's séances in which common individuals, many of whom were women, previously had acted as passive channels.[12] Spiritualism relied heavily upon women spirit mediums, debunking assertions deriving from the cult of true womanhood that women were suited to domestic roles and inclined to religious piety but not meant for leadership roles within spiritual spheres. Blavatsky radically diverged from prescribed images of women mediums as she was an active and powerful clairvoyant who could command spirits at will.[13]

Blavatsky's extraordinary demonstrations at Chittenden provoked accusations of fraud among participants there. However, Colonel Olcott found her displays intriguing and continued to share in séances and salons in Blavatsky's New York apartment. Out of these salons grew their own society for occult research, the Theosophical Society, which they cofounded in September 1878.

The Theosophical Society expanded rapidly in the late nineteenth century, with early membership increasing into the thousands and including distinguished converts such as inventor Thomas Edison. By 1880, the Theosophical Society boasted one hundred and twenty-one "lodges" chartered through the headquarters that moved from New York City to Adyar, India. As lodges' membership grew, they were reorganized into national units with ruling councils. The American branch, formed in 1886, was the first independent national section. It was followed by England (1888), India (1891), Australia and

Sweden (1895), New Zealand (1896), the Netherlands (1897), and France (1899).[14]

Frank Knothe joined the American Section of the Theosophical Society in 1896, rising to the rank of Assistant General Secretary by 1902. Knothe's parents became so deeply entrenched in theosophy that they named their first child "Alexander Fullerton Knothe" after the General Secretary under whom Frank presided. They gave their son the unfortunate nickname of "Alexander the Less" in deference to the General Secretary.

Upon Helen's birth in 1904, Frank and Maria believed, along with other Theosophical Society members, that their daughter was the reincarnation of Piet Meuleman. During visits to the headquarters of the Theosophical Society in Holland, Frank and Maria Knothe had befriended Meuleman, who was revered as "the mother of Dutch theosophy": "No member in any part of the world has worked more zealously and unselfishly for the Society than she," wrote Colonel Olcott of Mueleman.[15] The Knothes were said to have had a picture taken with Mueleman standing between them, their arms wrapped around her shoulders. Maria Knothe apparently remarked, "This feels like you're our child," to which Meuleman responded, "Forthcoming events oftentimes cast a long shadow." Frank and Maria Knothe held that their daughter Helen Knothe was the fulfillment of this "forthcoming event" prophecy and Mueleman's reincarnation. HK later would claim that her astrological chart was nearly identical to Mueleman's, though they were born in different months. She relished this exceptional, supernatural ancestry, keeping numerous photographs and books on Mueleman.[16]

HK became the most devout follower of theosophy in her family and a lifelong spiritual seeker. By age thirteen, she learned to read people's palms, studying the lengths of their fingers and the depths and position of the lines on their hands for insight into their character. She analyzed handwriting and collected autographs, believing both were likewise revealing.[17] HK later communed with fairies and would become a dowser who skillfully located water sources through a divining rod, as well as an inveterate practitioner of the Ouija board.

As a child, HK was also rather "bookish," passing much of her time reading and playing the violin. In the summer of 1917, her parents sent her to Camp Newfound in Bristol, New Hampshire, perhaps to wean her from her solitary habits.[18] Founded by Christian Scientists, Camp Newfound offered HK her first immersion in a spiritually focused, rustic setting. HK described summer camp as the beginning of her life as an individual apart from her family. She claimed to have thrived while canoeing alone, preferring solitude within the natural surroundings to groups of giggling girls. With an early proclivity for carefully documenting her life, HK constructed a Camp Newfound photograph album with images of picturesque Newfound Lake, as well as photographs of girls in bathing caps on docks casually posing together with Helen, all appearing to revel in their camp experience. HK also constructed a photograph album of a subsequent summer camp, Camp Winnisquam, which she attended in August 1919, and again included candid photographs, as well as images of the elegantly rustic camp, particularly its wood-paneled main lodge. HK's camp experiences were early introductions to the serene, spiritually rich, idyllic natural settings to which she would gravitate most of her life.

In 1921, Helen Knothe graduated from Ridgewood High School. Her parents suggested that she might attend college or a music conservatory in the United States, or perhaps study the violin in Europe. HK described herself as having "gypsy blood" and claimed to have chosen without hesitation to study the violin in Europe. She sailed for Holland with her mother on July 2, 1921.[19]

HK's choice was not only indicative of her own wanderlust, but also emblematic of a growing trend among a generation of "New Women" who came into their own in the years before and after World War I. These young women sought self-fulfillment rather than social service. They tended to be flamboyant in their pursuit of the arts and in their rejection of sexual conventions. They moved comfortably within bohemian circles in New York and urban centers of Europe, liberated from intense familial supervision.[20] Fashion styles introduced by the independent New Woman allowed greater freedom of movement and exposure of extremities. HK described her "little blue suit" in an August 15, 1923, diary entry as having shockingly "short trousers." She shared the New Woman's social impulses and carefree spirit, particularly once she was an ocean's distance from her father's stern reproaches and her mother's moderating influence.[21]

After her mother's departure from Europe, HK was to be chaperoned in Amsterdam by her maternal aunt Cornelia Dijkgraaf. They would live together in Dijkgraaf's apartment in the Theosophical Headquarters established by Piet Mueleman.[22] The Knothe family maintained ties to the Theosophical Society despite the many controversies surrounding theosophy. In the midst of the Theosophical Society's late-nineteenth-century rapid expansion, the Society for Psychical Research in London had filed an 1885 report debunking Blavatsky's "fraudulent phenomena," in particular her claims regarding visits and letters from the brotherhood of the Masters.[23] The Society for Psychical Research threatened suit against Blavatsky. She, in turn, resigned as corresponding secretary of the Theosophical Society and left the headquarters in India.[24] Scandals and accusations plagued the Theosophical Society for the next three decades but, remarkably, did not deter its progress.

In 1909, Frank Knothe had had a falling out with the Theosophical Society and its president, Annie Besant. He temporarily resigned his membership, writing a scathing letter to the Inter-State Branch regarding Besant's inept, domineering leadership style and her "inexplicable" defense of Charles W. Leadbeater. In one of many Theosophical Society imbroglios, Leadbeater—a principal advisor to Besant—had been accused of inappropriate sexual advances toward young men in the Society. Frank Knothe later rejoined the Society, but he and his wife did not immerse themselves to the extent that they had prior to the Leadbeater affair.

Shortly after HK and her mother arrived in Amsterdam, they traveled to Paris to attend a theosophy convention where Jiddu Krishnamurti would speak. In 1909, the Theosophical Society, under the direction of Annie Besant, had chosen Jiddu Krishnamurti—an undernourished, fifteen-year-old Indian boy—as its spiritual leader.[25] Krishnamurti's father was a widower with nine other children who worked as a day laborer at the Adyar headquarters. Charles W. Leadbeater conducted a series of interviews with Krishnamurti, after which he declared Krishna to be the spiritual guru of the Theosophical Society, the body chosen as the vehicle for the coming messiah. Besant became legal guardian of Krishna and his brother Nitya and Leadbeater their primary teacher in Adyar. In 1911, Krishna and Nitya were sent to England where they continued their theosophical education over the next ten years.

Krishnamurti was an unskilled, unmotivated student, but he had a natural charisma and hypnotic rhetorical style that proved inspirational to crowds of believers. His address in Paris that HK and her mother attended was Helen's first introduction to "the shy young man." She later wrote: "He was slight, of average height and, being Indian, his coloring was dark. His hair was black and glossy, his features were classic, with aquiline nose, great fringed eyes, and a sensitive mouth."[26] HK immediately found him attractive, which was not uncommon, as Krishna was known for his particular allure among young followers.

The interwar years were a time of burgeoning youth movements as a generation of young people, in the wake of the devastating first World War, searched for alternative ideas and charismatic leaders like Krishnamurti. Modernist youth were drawn to pacifism, humanitarianism, and international idealism, all of which Krishnamurti espoused. Historian Peter Washington described the Theosophical Society between 1919 and 1928 as "a sort of junior league of nations" in its widespread appeal to young people.[27]

HK's Aunt Cornelia worked as the national representative for the

Theosophical Society's Star Branch or the Order of the Star of the East, a subset within theosophy comprised of mostly young Society members. Thus HK, upon her arrival in Amsterdam, immediately was caught up in this spiritual youth movement. She officially joined the Star Branch in October 1921, and, in conjunction with her violin studies, immersed herself in European theosophy. She read theosophical texts such as *At the Feet of the Master* and articles on the "Eastern School of Theosophy," forging an increasingly deep spiritual life steeped in occultism and notions of right discipleship.

Theosophical discipleship was characterized by suppression of one's ego in love and obedience toward the supernatural Masters, as well as toward earth-bound teachers such as Krishnamurti, Annie Besant, and Charles Leadbeater. HK wrote in her 1923 diary, "I will learn too the pure and Holy Way of loving and purge myself of all selfish and personal traits."[28] Theosophical Society leaders urged disciples to become "initiates," enlightened followers adept at meditation and clairvoyance, while also selfless in their devotion to their teachers. This early indoctrination would shape HK's spiritual journey as she remained throughout adulthood committed to living purposefully and deferent to guides whom she saw as more "developed" than herself.[29]

HK's involvement in theosophy became particularly complex in September 1921, when Jiddu Krishnamurti visited Amsterdam and claimed to fall in love with the seventeen-year-old.[30] He wrote to her a few months later as he and Nitya prepared to leave for India:[31]

Amsterdam: Nov 14:21

Darling Helen,
  I love you with all my heart and soul and shall always do so.

Bless you,
Krishna[32]

Krishnamurti's overtures intrigued HK, but her romance with Krishna did not at first distract her from other relationships, such as

a growing friendship with Amsterdam Theosophical Society companion Koos van der Leuw, or from her violin work.

In Amsterdam, HK studied the violin with Louis Zimmerman,
concertmaster of Willem Mengelberg's Concertgebouw Orchestra and
a renowned violin teacher in Holland. She was devoted to practicing
and attending concerts, immersing herself in the symphonies of
Mozart, Mahler, Bach, Beethoven, and Brahms. In the winter of 1922,
HK became deeply inspired by the talent and interpretation of Irish
violinist and composer Mary Dickenson-Auner. HK arranged to
study with Dickenson-Auner in Vienna in February of 1923.

Helen's parents, however, had begun to suggest that she return to
the United States. One year prior, in a February 1922 birthday letter,
Frank Knothe had included a check for $1,025.90 for three months'
support but also suggested that Helen, as she turned eighteen, should
begin to handle her own affairs. In a letter a few months later, he alluded to financial difficulties at home and proposed that HK return
and take courses at the Ridgewood secretarial school. He contended
that Helen should consider the "material," not "dream away our earth
life."[33] Her mother, however, felt that Helen should continue her musical studies with Zimmerman.

By December, the family's financial troubles had passed, but her
father remained opposed to HK's staying in Europe. Frank Knothe
wrote, "Really, you make my head swim with your wanderlust."[34] In a
letter written at the same time, her mother began to encourage Helen
to return to a university in the States. "I am so glad you seem to like
the Radcliffe or some American college plan," wrote Maria Knothe.[35]
Her parents' concern would have stemmed in part from HK's deepening immersion in the Theosophical Society as HK grew increasingly close to both Krishnamurti and president Annie Besant.

Despite her parents' entreaties, HK resolved to remain in Europe.
Photographs from that time depict her in school-girl attire posing
with Krishnamurti on a tennis court or running foot races or leaning
against the fender of a sporty roadster. It was a lighthearted time for

HK, despite the conflicting pressure of her parents' appeals to return home and Krishnamurti's and Annie Besant's petitions for her to stay. In a January 2, 1923, diary entry, HK questioned "how much family ties should influence one in making decisions for [one's] self." In the next day's entry, she described Krishna's love for her, as well as Besant's "urging [her] to go to Australia" to study at a Theosophical Society institute there, to which Knothe wrote defiantly, "I love Krishna and Annie Besant! *Will* go!"[36]

Such youthful, defiant emotions permeated HK's 1923 diary. One day, she was dissatisfied with a dressmaker who was late, another she depicted as a "heavenly day" in which she practiced the violin for two hours and received a love letter from Krishna and "just sat and loved him." She also described various flirtations, regretting having "dressed so young" for an appointment with a handsome doctor.[37] It was a privileged, blissful time for the idealistic musician and spiritual seeker, "What a wonderful life, I'm tingling with it," she wrote.[38]

HK also began to deepen her attachment to her natural environment at this time. She had undergone minor abdominal surgery in Amsterdam in January 1923, and suffered postoperative symptoms of sleeplessness, aches, and excessive fatigue.[39] In late March, a Viennese doctor diagnosed her as being severely anemic and prescribed radium rays, a treatment dating back to Marie Curie that later was proven to cause both anemia and cancer.[40] Knothe described the radium rays as "agonizing." The doctor modified her treatment, using quartz and helium lights instead, along with homeopathic medicines. HK eschewed taking the medications but believed in the curative powers of the rays, despite painful skin burns. From that time forward, she would herald the healing strength of light and, in particular, of the sun's rays.[41]

HK also repeatedly retreated to the woods of Austria, finding peace alone beneath a tree or on a mountain. She described solace in natural settings: "I lay off in the sun by myself under the pine trees and sang to myself" or "I took a solitary ramble up the mountain with my knitting."[42] She referred throughout her 1923 diary to her penchant for solitude, particularly in nature. She chastised herself for her inability to be more collegial, claiming "This social life does *not* suit me" and "I must learn to mix." She wrote later in the summer, "I do try to keep clean and pure, but it's the devil to get rid of Helen in my Kosmos."[43]

Theosophy's emphasis upon suppression of one's ego instilled in HK conflicted feelings about self-worth, about her many privileges, and about Krishnamurti's affections for her. She wrote in July, "Oh, he loves me, God. God, why? I must be worthy, I *will* be worthy!" and in August, "Oh, God, that I may quickly prove my worth!"[44] HK also expressed a desire to serve the supernatural Masters as well, imploring "My Master, may I be worthy of your trust in me." She expressed a youthful, self-critical insecurity, describing her "legs as the fattest and biggest part of me" and writing regarding Krishnamurti, "pray God he loves me still and overlooks my many failings." Yet she also intimated a sense of her unique gifts, particularly in her search for mastery of the violin, "Where do I keep what I've got. Why can't I use it?"[45]

As a young, self-critical seeker, HK was particularly open and vulnerable to the persuasions of various authority figures. Theosophical Society president Annie Besant wrote to her in February 1923, regarding her parents' request to return to Ridgewood, "I suppose you must go if your parents insist; but it is a great pity." Besant encouraged HK to continue to develop her musical talents as "the fiddle" might "make [her] independent" from her family.[46] As Krishnamurti's primary love interest and the reincarnation of Mueleman, HK had gained significant stature within the Theosophical Society. Nonetheless, her position as an "independent" New Woman and devotee of theosophy and music was financed by her parents and sustained by her emotional and spiritual dependence upon Besant and Krishnamurti.

HK moved to Vienna in 1923 not only to study with Dickenson-Auner, but also to help prepare for an upcoming Theosophical Society congress there.[47] The international theosophical congress met in

Vienna at the end of July, and Krishna and Nitya attended, along with a host of Theosophical Society and Order of the Star of the East members. According to a July 22 diary entry, Krishna was "cruel and harsh" to Helen upon arrival, but still enamored of her.

At the end of the Vienna congress, HK left her studies with Dickenson-Auner and joined Krishna, Nitya, and their companions at Villa Sonnblick in the village of Ehrwald high in the Austrian Tirol. The villa's chalet-style main structure later would inspire HK's designs for her stone houses in Vermont and Maine. She wrote regarding her decision to leave her violin studies:

> Do I foolishly put "occult development" before music? I don't put it there, it's just all my life to me. There is nothing else in the world. I couldn't but believe in the Masters and want to be like them.[48]

She also described the extent to which she was persuaded by Krishnamurti to join his entourage:

> He inadvertently made me question the rights and wrongs of Father's decision for me to carry on with the fiddle. One cannot serve Mammon and God at the same time he says. Still *is* not the fiddle the apparent Dharma set before me? There are many 4th rate fiddlers in the world and so *few* initiates, he says.[49]

HK brought her violin with her to Ehrwald and claimed, "I do love my fiddle now and would miss it if I did not have it to practice on."[50] However, her Theosophical Society discipleship would take precedence over her music.

Beyond his romantic overtures and her spiritual obligations, the appeal for HK to join Krishnamurti's inner circle would have been great. Krishna surrounded himself with close friends as a buffer against legions of devoted followers and mainstream critics. HK found this group both spiritually enriching and oddly glamorous. Krishnamurti and his brother Nitya, despite their modest beginnings and Krishna's disparagement of riches, led privileged, wealthy

lives, traveling extensively and residing in extraordinarily picturesque places. A wealthy, generous American woman and member of
the Theosophical Society, Miss Dodge, had granted Krishna and
Nitya lifelong annuities—five hundred pounds annually for Krishna
and three hundred for Nitya. In turn, the brothers wore garish, cus-
tom-made suits, shirts, and shoes; gray spats and gray homburg hats;
carried gold-headed canes; and drove sporty automobiles.[51]

Other wealthy patrons offered Krishnamurti unrestricted use of
their estates. In Holland, the Baron van Pallandt turned over his
grand, early-eighteenth-century ancestral Castle Erde and its five
thousand acres to Krishnamurti. The chalet Villa Sonnblick in Austria, where HK spent August and September of 1923, was available to
them through the generosity of a wealthy friend of one of Krishna's
tutors at Adyar. Theosophical Society rhetoric espoused austerity
and minimalism, parroted by HK's self-reproach, "I can't play both
high and low in life." She was taught that she could not enjoy wealth
and its advantages at the same time that she sought selfless detach-
ment from the material world.[52] She claimed, "I want simplicity to be
my forte."[53] Yet Theosophical Society leaders and members typically,
like HK, were entrenched in privilege.

HK's infatuation with Krishnamurti deepened at Villa Sonnblick.
Mary Lutyens, a member of the party that August, described Krishna
as "extraordinarily beautiful," imbued with a "personal magnetism"
that made it "inevitable that many women should fall in love with
him." However, as the body for the coming messiah, he was supposed
to have sublimated sexual love for spiritual pursuits. Lutyens described
other women as teeming with jealousy and "heart-burning" as they
recognized that Krishna clearly had begun to favor Helen Knothe. Ruth
Roberts, also a member of the group, had had "a recent flirtation" with
Krishna in Sydney, Australia; two years later, Mary Lutyens herself as-
serted his affections were directed to her; and Rosalind Williams would
be a later love of Krishna's. Knothe in her 1923 diary asked, "why me?"
claiming that Krishna could have chosen to be with any woman there.[54]

Interpersonal conflicts within this close circle and Krishna's multiple love interests spoke to the youthfulness of the group. In 1923, Helen was nineteen years old; Krishna was twenty-seven. Her photographs from Villa Sonnblick depicted attractive groups seated cross-legged in the grass, women in loose-fitting, Indian print dresses, all with ebullient smiles. They were happy, privileged young adults, preoccupied with romance, games of British rounders, and hikes in the nearby Alps. They were also at the center of a growing religious movement. By the late 1920s, the Order of the Star in the East reached its apex in membership at thirty thousand, and the larger Theosophical Society claimed approximately forty-five thousand members.[55] Helen and Krishna's intimate relationship from 1923 to 1927 coincided with this expansion as HK was involved deeply in theosophy at the height of its influence.

HK's role within this movement was elevated dramatically on August 13, 1923, as she became the primary spiritual channel for Krishnamurti. Krishna began to move in and out of consciousness, experiencing excruciating physical pain attributed to his beginning "the process," his body's preparation for his work as the spiritual vehicle of the coming messiah. Historian Peter Washington described these episodes as wholly psychosomatic and recurring during stressful periods throughout Krishna's life. Madame Blavatsky and other theosophists reported similar episodes of pain, but, according to Leadbeater, none had experienced Krishna's physical misery.[56] Annie Besant described an early episode of the process and HK's role in it:

> At dinner time, he was obviously hardly conscious and almost directly afterwards he went right "off" and the body began to sob and groan. We all sat very quietly outside except the faithful Nitya—who presumably sent for Helen as he thought she might help him. It lasted until nine o'clock, when he came round and went off to bed. But at twelve o'clock he began again and again Helen sat by him till one o'clock and once again in the early morning. He said that Helen was very nervous and naturally so as at first it is awful to witness such suf-

fering and to realize his consciousness is not there. It is very curious as he seems to need a woman's presence and also that the vitality of Americans seems to supply something that he needs.[57]

HK described his agony during the episodes, "He had the same terrifying pain as yesterday, 'like a red hot poker stuck in a raw wound.'" HK and Nitya attended Krishna nightly for over a month as he went through this process.[58]

Annie Besant described the awkwardness and significance of HK's role in Krishna's process: "It is a curious experience for a young girl," wrote Besant, to attend to her religious leader—and boyfriend—who continually mistook her for his mother during painful fits of delirium. Charles Leadbeater suggested that HK was a medium through whom Krishna's mother could help him endure the suffering, and as such, HK's spiritual stature became formidable within the group as the channel for Krishnamurti on his path to becoming the theosophical messiah.

HK felt that it was "fair play" for her to comfort Krishnamurti, as he had been rubbing and healing her spreading warts, but the extent of HK's care far surpassed Krishna's. HK claimed to receive messages from the Masters during each episode.[59] She referred to Krishna as becoming a "sweet boy" during his periods of delirium, and his turning to her as a mother or a sister.[60] However, when he "went off," HK's role during the process also became increasingly amorous. HK described embarrassment about what Nitya must have thought: "I wonder what Nitya thinks of while all this is going on, while he lies in bed and hears his brother and a strange woman carry on."[61] Their passionate exchanges escalated to the point that HK worried "what a dirty astral body I must have." She described kissing Krishna and their sensual spots being under her arms or on his right eye.[62] HK did not divulge in her diary whether they actually were having sex, but she claimed, "many things I can't even write."[63]

On September 20, 1923, HK traveled with Krishna and his entourage to Castle Erde in Holland as his process continued. She left briefly for Vienna, where she spent one week studying again with Dickenson-Auner. The brief stint away from Krishnamurti, and the rigorous violin session, solidified HK's desire to heed her parents' request and return to the United States. She would forego her European violin studies and separate briefly from the demands of her Theosophical Society entourage. Krishnamurti and Nitya also would leave Europe and travel to their cottage in Ojai, California, lent to them as a place for Nitya to convalesce after he was diagnosed with consumption in May 1921. The brothers hoped that the warm California air might relieve Krishna's pains.

Krishnamurti asked HK to join them in California, but she felt that she could go with the brothers only as far as New York. In preparation for their journey, HK recognized her public position as Krishamurti's travel companion, reflecting in her diary, "I'll be on the boat with K and will need something smart."[64] Such a comment denoted conventional notions of femininity, more laden with Ridgewood values than spiritual ones, as well as a sense of peripheral celebrity in the presence of her teacher.

HK's physical, romantic role in Krishna's process had ceased temporarily on October 18, much to her relief. She wrote, "I felt I *didn't* want it and if he didn't need it, why there'd be an end to that part of the business . . . Before, during and after, both felt cleaner!"[65] Yet they traveled on the steamship to the United States without a chaperone and lapsed back into their amorous relationship. HK recounted sneers from the cabin boys as she moved from her next-door cabin into Krishna and Nitya's.[66] She wrote in her diary, "if they had an inkling of what went on!"[67] HK described being in Krishna's bed as well as Nitya's and noted Krishna's anger over her closeness to Nitya, along with her own duplicitous feelings. "It was a weird experience dividing myself between the two brothers," she reflected after having been romantic with one and then the other. She wondered what in the world her family would think.[68]

Krishnamurti's process continued in Ojai, where he was said to have called out for Helen repeatedly.[69] Lady Emily Lutyens—an active British Theosophical Society member, mother of Mary, and part of Krishna's inner circle—wrote imploring letters to HK asking her to return to the group:

> Are you coming to Europe this summer? Oh! how I hope so. I want to see you so badly. I have got such lovely plans for the autumn and wonder if you will only come. Do adopt me as a mother. I am longing to adopt you as a daughter "for keeps"! I have been thinking so much of you. I love you very much, and I hunger to take you in my arms and give you a big hug. You *must* come.[70]

Lutyens wrote a few days later: "only you can decide—of course I feel that there is only one thing now for you to do and that is come to England and see Mrs. Besant, but I cannot urge this." She added, "I love you dearly—my shoulder is aching for you to come and weep on it." Lutyens described a silk scarf that she had purchased for Helen, signing the letter, "Mother want to be!"[71] Lutyens's entreaties highlight Theosophical Society members' seducing HK to distance herself from her family and to redirect her devotions to the surrogate Theosophical Society and Krishnamurti.

HK spent the fall of 1923 and the ensuing spring living at Cottage Place with her family. She struggled to conform to the social mores of Ridgewood, and she gradually grew apart from her father, in particular. She described feeling "numb and lifeless" in New Jersey, and being courted by suitors such as Bruce Ellison who lacked Krishnamurti's "idealistic purpose in life."[72] In early November, she wrote that "the least I could do was stay here and make Father happy," but a few weeks later depicted "grand family rows" regarding her father's insistence that she stay in Ridgewood and learn to be "practical."[73] Frank Knothe criticized Helen for "growing away from him," and Helen lamented their falling out: "Poor Pop, he and I grow farther away day by day, and I always thought we'd be such chums, it's too bad."[74]

Despite their bohemian leanings, the Knothe family seemed conventional relative to HK's recent consorts. HK described immediately feeling "at home" at a Theosophical Society lodge meeting in New Jersey and wrote, "Here's where I belong." In her diary, she reflected upon letters from Krishnamurti in which he expressed fear that she was "living the frivolous life . . . forgetting him and the Masters and [her] purpose in life."[75] She, too, felt that her Ridgewood days lacked purpose and mused about her craving to be with Theosophical Society members:

> I wonder just what is the meaning back of my craving, yes, craving for proximity to developed people of this type. Am I a leech that I want to be seen in their company or do I recognize their greatness and aspire to it? Then why not stand on my own legs among my equals and inferiors?[76]

HK recognized her attraction to leaders or teachers, to "developed people," but also questioned her motivations. Would she aspire to their greatness or simply bask in the glow of their success? Would she stand on her own?

HK determined that she would not remain in Ridgewood and wrote, "I'll be damned if I'll be stuck in this mud hole with these tadpoles while all the big fish swim by."[77] She bid farewell to 1923 by sneaking out of a New Year's Eve party and retreating to the woods alone. In the quiet forest, she "nearly heard the rustling of wings as '23 flew sighing past." She described looking ahead more than behind but expressed dissatisfaction with the outcome of the year that "has as yet shown no results, causes me more pain than joy."[78]

HK wrote in a later notebook that her mother helped her return to Europe the next summer, while her "distressed" father would not even say goodbye.[79] On June 15, 1924, Helen met Krishnamurti and Nitya in New York and traveled with them to England and then on to Holland for a Theosophical Society and Star Congress. Following the congress, Helen, Krishna, Nitya and their familiar entourage participated in a Theosophical Society camp at Ommen on the elegant grounds of Castle Erde.

Ommen Camps or Star Camps were an annual Theosophical Society event throughout the 1920s. Krishnamurti presided over them, and thousands attended, mostly young people who sat beneath trees and slept in tents on the castle grounds. The Theosophical Society elite—of which HK was a part—stayed in castle chambers or in specially constructed private huts. The Ommen Camps offered uplifting meditations and lectures about religious potential and political idealism, resonant with postwar searches for self-improvement and alternative living. Their spiritual monarch, Krishnamurti, entranced the crowds of seekers.[80]

After the 1924 camp, on August 18, HK traveled with Krishna and the others to the Dolomite Mountains. There they stayed in an eleventh-century castle perched like a Tibetan monastery above the Italian village of Pergine. Helen and Krishna began to drift apart in Italy, as they spent little time alone together, except during his "process" episodes when, as HK described, she held "him not as a lover but as a mother." HK participated in daily lectures that Krishna delivered to women in the group under the shade of an apple tree. He talked of preparing for the coming Lord, insisting that the women not pursue marriage or a home of their own as this would not allow them to serve the Lord when he came. HK described his having "mercilessly high" standards. Mary Lutyens recalled, "He was very hard on us at Pergine, often making us cry by the home truths he told us."[81]

Despite Krishna's overbearing teachings and her diminishing role in his life, HK described Pergine as a holiday and a carefree time. The group brought its own Austrian vegetarian cook who served elaborate meals in a vast dining room. HK practiced the violin in the castle dungeon, played volleyball and rounders, and explored the surrounding area. Her only concern seemed to be Nitya's failing health. Moreover, at Pergine, HK deepened her commitment to theosophy as she agreed to the "lovely plans for the autumn" to which Lady Emily had referred. She would travel to Charles W. Leadbeater's Manor House, a Theosophical Society school in Sydney, Australia, where she would be brought further along her path of theosophical discipleship. Krishna and Nitya's patron, Miss Dodge, would pay her fare.

On November 2, 1924, against her parents' wishes, HK sailed from Italy to Bombay (now Mumbai)—along with Ruth Roberts, Betty and Mary Lutyens, Krishna, Nitya, and Lady Emily as their chaperone—en route to Australia. HK described her first experiences of India as revelatory. She was both overwhelmed by Indian beauty and luxury and stunned by the want and poverty. She noted bizarre funeral marches with burning goats, majestic palm trees, and meals eaten off of banana leaves.[82] The group traveled to the "Home of the Masters," the grand Theosophical Society Headquarters in Adyar, Madras (now Chennai), on the southeastern coast of India. Having left schoolgirl clothes far behind, HK wore Indian tunic dresses to the Theosophical Society ceremonies there. Krishnamurti donned traditional Indian garb as he processed to the headquarter's lotus pool and the voluminous banyan tree under which his followers, including HK, gathered.

HK remained in Adyar until December 1924, at which time she continued her journey to Australia, beginning what she described as "the most flowerlike, felicitous time of [her] life."[83] She moved into the fifty-five-room Manor House, a Theosophical Society commune overlooking Sydney Harbor, where the exceedingly controversial Charles W. Leadbeater presided. Krishnamurti had written to Leadbeater in reference to the talent he saw in HK:

> With regard to Helen, I have been urging and contriving to facilitate her going to you. That has been my dream, ever since I met her, and I have been trying to prepare her to take full advantage of her being with you. I think she is somebody and will be of use later and for that I have done everything to make it possible for her to go to you.[84]

Krishna's insights proved prescient as HK thrived at the Manor, relishing in the school's rigorous meditation, yoga, and study. She also played the violin frequently, as well as the pipe organ, and did secretarial work for Leadbeater, honing her typing skills.

HK furthermore fulfilled her foremost task of progressing as a disciple, primarily through meditation and yoga, in order to take important occult steps along her spiritual path. She later described having prepared like a nun or priest for confirmation in "most holy service to Bishop CWL." On May 8, 1925, at 1:43 A.M. of Wesak (an occult festival celebrated on the first full moon in May), "Helen was accepted." She advanced to one step below "Initiate," which Lady Emily reached on that Wesak, and moved one step above "probation," which Mary and her sister Betty earned. The persistent post-Wesak question around the Manor was "How far on are you?" to which HK would have been proud to respond "accepted."[85]

In the fall of that same year, Nitya died of consumption. A few days after his death, HK, Ruth Roberts, Mary and Betty Lutyens, and Leadbeater met Krishnamurti in Colombo, Ceylon (now Sri Lanka), and then traveled together to Adyar for Nitya's funeral. HK described her relationship with Krishna as having been further altered by Nitya's death: "It changed [Krishna]. I was still with him but differently."[86]

HK and the group remained in India for a December 28, 1925, Star Congress and Theosophical Society convention. There, Krishnamurti underwent a radical transformation while sitting under the Headquarter's banyan tree, addressing a crowd of over three thousand people. As he was describing the anticipated World Teacher, Krishna made a linguistic and psychic shift to the first person. He began by saying, "He comes only to those who want, who desire, who long." Krishnamurti then changed his facial expression and said, "I come to those who want sympathy, who want happiness . . . I come not to destroy but to build." Followers, including HK, believed that at this moment Krishna became the World Teacher, marking a spiritual triumph for the Theosophical Society as its leader spontaneously transformed into the messiah. This event, which they presumably had anticipated, ironically ushered in a period of conflict and confusion within the Society. Krishnamurti would spend the next several years mired in doctrinal and administrative disputes with President Annie Besant and other aging leaders, whose past teachings were threatened by his sudden increased power and autonomy.[87]

At the end of January 1926, HK returned to Sydney, where she, too, found herself embroiled in a conflict related to Krishna. Rumors spread in newspapers that HK and Krishnamurti were engaged to be married, which Krishna emphatically denied. He claimed, "Any report regarding the engagement is absurd. It is really too terrible." HK likewise confirmed that they had not been engaged, but this controversy solidified Krishna's loss of interest in her. HK wrote later to Mary Lutyens, "I was deemed superfluous and dropped."[88] HK did not receive another love letter from Krishnamurti after 1926.

In 1927, HK culminated her studies in Sydney and returned to Ridgewood, feeling abandoned by Krishna and disillusioned by the Theosophical Society and its persistent quarrels. Her temporary departure from the flailing Theosophical Society proved wise as Annie Besant and Krishnamurti continued to clash over doctrinal issues. In August 1929, Krishna announced at an Ommen Camp that "truth is a pathless land," disclaiming the fundamental Theosophical Society teaching that a prescribed path is essential for spiritual growth. He renounced his role as the coming messiah, dissolved the Star Branch, and officially resigned from the Theosophical Society. Two years later, Krishnamurti experienced a complete memory loss—though close friends found the oblivion to be selective—and made a new spiritual life for himself in America, with his home base in Ojai, California.[89]

HK's six-year immersion in the Theosophical Society proved deeply formative as she continued throughout her life to cultivate a rich spirituality based upon theosophical teachings, supernatural phenomena, and personal discipline. Her experiences in early adulthood fostered a profound sense of independence; she was a New Woman traveling extensively and eschewing conventionality. She honed her skills as a violinist, moved to the peripheral limelight of a growing religious movement, and cultivated a growing connection to natural landscapes. However, her time with Krishnamurti and his entourage accentuated HK's potential vulnerability as a young seeker highly susceptible to the persuasions of a charismatic leader. Tensions with her family would persist, taking on new iterations as her devotions shifted to Scott Nearing and his dogmatic austerity. HK continued in her pursuit of more "developed" people and in her search for a purposeful life, but also grappled with the question of whether she would "stand on her own."[90]

CHAPTER 2

Political Awakening

The Early Years with Scott Nearing ✀

WHEN HELEN KNOTHE moved from Australia back to her family's home in Ridgewood, she resumed violin practice and worked as a part-time secretary for her father. HK later described her return to the States in a journal entry as, "Back home. Not happy."[1] However, this mundane stint at home was short-lived. In 1928, HK's father asked her to enlist Scott Nearing as a speaker for a club meeting at their local Unitarian Church. HK had met Nearing at a political gathering at his Ridgewood home seven years prior.

Nearing was known for his skills as an orator and for his outspokenness against capitalism. He had begun his career as an economics professor at the University of Pennsylvania (1908 to 1915), where he heralded the political and social tenets of Progressivism and forcefully denounced the exploitation of child labor and profiteering through wars.

In 1908, Nearing had married Nellie M. Seeds, a graduate of Bryn Mawr who earned her masters of liberal arts from the University of Pennsylvania in 1910 and her doctorate in education in 1915. Seeds shared her husband's intellectual commitments as well as his penchant for reform. In 1912, they co-authored *Woman and Social Progress: A Discussion of the Biologic, Domestic, Industrial and Social Responsibilities of American Women*, in which they argued that industrialization redefined women's work and public roles such that women had "the opportunity to contribute in many different spheres, their share toward Social Progress."[2] That same year, the Nearings' first son John had been born, and they adopted a second son, Robert ("Bob"), two years later.

In his teaching at the University of Pennsylvania, Nearing called for moral regeneration in American society, espousing social religion and pragmatic economics in lieu of capitalism and its consequent cultural decay. In 1915, he was fired from the University of Pennsylvania for his condemnation of American big business and, in particular, its former abuse of child labor.

Nearing took a position in the economics department at the University of Toledo, which he subsequently lost after the publication of his controversial 1917 pamphlet, "The Great Madness." Nearing argued in his pamphlet that the principal aims of war were commercial. He was not only fired from the University of Toledo but, under the Espionage and Sedition Acts, indicted for ostensibly obstructing recruitment and enlistment in the armed forces. Nearing eloquently spoke in his own defense and was not convicted, but he was blacklisted from academia from that time forward.[3]

Nearing fell back upon his oratory skills, later claiming that between 1915 and 1935 he lectured publicly "eight or ten times in a week—in all some four hundred lectures a year."[4] He regularly participated in public debates, perhaps his best known being a 1917 debate with Clarence Darrow on the topic "Will Democracy Cure the Ills of the World?" at the Chicago Workers' University Society. He also debated Bertrand Russell on "Bolshevism and the West: A Debate on the Resolution 'That the Soviet form of government is applicable to Western civilization'" in London in 1924.

Nearing tried his hand at politics, running unsuccessfully for Congress on the Socialist ticket in Manhattan in 1918 against Fiorello La Guardia. His losing platform called for opposition to war and an end to American imperialism. He ran a promising campaign for governor of New Jersey on the Communist ticket in 1928, but lost that election as well.

In the early 1920s, Nearing taught a variety of courses at the alternative Rand School of Social Sciences in New York City. The Rand School was organized by the Socialist Party (of which Nearing was a member until 1923) in order to serve the educational needs of trade unions and the socialist movement in the United States.[5] Nearing shared an office with union organizer Eugene Debs, who had described him as "the greatest teacher in the United States."[6]

However, Nearing's wife Nellie Seeds Nearing would prove to have the more successful teaching career. She lived out their progressive edict for expanding women's work, serving as executive secretary, then director, of the Rand School of Social Science where Nearing and Debs taught. After she and Nearing separated in 1925, she worked for, and later became director of, the Manumit School in Pauling, New York. Founded by William Fincke as "a laboratory school of the American labor movement," the Manumit School was a prototype of the New Education movement and espoused freedom, cooperation, and growth of the individual student. Manumit offered a special tuition rate to children of trade unionists and boasted an interracial student population.[7] The school also operated a large dairy farm on which Nearing worked in the summers. Though separated, Nearing and Seeds would remain married until her death in late 1946, never officially divorcing.

In the mid-1920s, Nearing expanded his political and economic interests internationally. Traveling abroad for his research, he completed a pamphlet on the British general strike in 1925, a piece on Russian education the following year, and a pamphlet on China the next. By 1927, he had published over twenty-five books and pamphlets on national and international topics, remaining a prominent figure in the burgeoning left for most of the twenties, as well as a constant foil in leftist party politics.

Nearing resigned from the Socialist Party in 1923 because he believed the party "remained aloof from the rank and file" and was out of touch with the American worker. He also criticized the Socialist Party's insufficient connection to Russia: "Moscow is strong; the Workers Party is weak."[8] Nearing maintained a tenuous relationship with the Communist Party throughout the 1920s and was denied membership on several occasions.[9] A Communist Party critic of Nearing, Michael Gold, described Nearing as a "solo dancer," an individualist who "followed his own mystic impulses and logic," which "gums up a political movement where men must work together."[10] Nearing finally was accepted into the Communist Party in the late twenties, but resigned in 1930 as the Party simultaneously expelled him. He would claim to continue to uphold communist principles as a fellow traveler but would privilege his own "inner promptings" over the comradeship of the Party.[11]

Thus, in 1928, when Nearing became reacquainted with Helen Knothe, he had been banned for over a decade from mainstream academic institutions, had resigned from the Socialist Party, had lost two political campaigns, was separated from his wife and children, and was poised to leave the Communist Party.[12] Historian John Saltmarsh described their immediate connection with one another as providing a kind of spiritual "redemption" for the downtrodden economist.[13]

HK at age twenty-four was curious, energetic, and spiritually charged. She exuded a youthful elegance and sophistication, dressing stylishly and often holding one hand to her chest like an actress. She had a carefree way about her, whether sitting cross-legged in a modified yoga pose with her pervasive knitting or standing straight at five feet six inches tall.[14] She was also a seeker looking for a purposeful alternative to her life in Ridgewood.

Shortly after HK telephoned Nearing to ask him to speak at her father's club, he invited her to join him on an errand to upstate New York. HK agreed and claimed to have immediately become infatuated with Nearing on that drive when she learned that he was a vegetarian and that he believed in fairies. They kissed that day, marking the beginning of a fifty-five year intimate relationship.[15]

Helen Knothe's attraction to this radical socialist twenty years her senior was perhaps an extension of her propensity to seek out older, paternalistic teachers. She also would have been drawn to Nearing's radical eschewing of the social conventions with which she was struggling to conform. He deemed polite society a fraud, rejecting the use of bourgeois china and cutlery, preferring instead to use wooden bowls and chopsticks. Nearing spurned first-class steamship tickets on family trips, and HK likewise had expressed preference for the livelier atmosphere of second-class travel.[16] Furthermore, Nearing's Ridgewood house was an interesting, eclectic gathering place for leftists from New York City, including guests such as Margaret Sanger, Norman Thomas, and William Fincke.[17] He was a radical seeker whose mantra, "*semper major*," always more, and his condemnation of any profligate use of time would have attracted a consummate, rigorous seeker such as HK.

Their early relationship proved deeply romantic. Nearing wrote to HK on November 11, 1929, "Only last night I was thinking how we had developed the technique of getting on together. I never was so close to anyone else."[18] He addressed letters to her as "Dearest Best," "Best Beloved," and "Sweetheart," and might conclude with "Be Well and Strong, dear love."[19] Helen wrote in a later diary after Scott brought her flowers, "I appreciate his nice qualities so much." She described him affectionately, "He is a treasure."[20]

However, their relationship could be rife with tension, particularly with regard to the fact that Nearing remained married to Seeds. Nearing continued to work on the grounds and in the garden at the Manumit School and stayed in close contact with his sons. Perhaps he did not divorce Seeds because they shared intellectual, social, and political commitments, as well as long-term, mutual regard. A later Vermont neighbor claimed that Nellie Seeds never would grant Nearing a divorce because of his persistent philandering. She was reluctant to have another woman suffer the infidelities that she had.[21]

Nearing's sexual affairs with other women did indeed prove to be a source of conflict with HK. Nearing wrote to HK on December 20, 1929, that his former secretary in Toledo, Grace Small, claimed to be three months pregnant with his child. Though Nearing denied paternity, he admitted to having had intimate relations with Small recently. He would have many such liaisons with other women in the future.[22]

HK and Nearing's early correspondences demonstrated an ambivalence regarding whether they were friends or lovers, as well as uncertainty as to their future together. There was a bohemian openness to their relationship as HK continued close friendships with suitors such as Bruce Ellison, who would move to Vermont to be close to her in the thirties.[23] In letter exchanges in which Helen mentioned marriage to Scott, he callously suggested that, if that was her inclination, she should consider one of her other suitors instead.[24]

Frequent disagreements peppered their first decade together. After a particularly heated exchange, HK would declare, "We'll have to carry on alone" or "bad to fight so." However, most flare-ups depicted in diary entries would be followed the next day by an entry such as "friends again." HK wrote upon Nearing's return from a trip "a joy to see him" and described Nearing, who romantically would bring her daffodils and ice cream, as the "nicest person I've met." They read *Patience Worth, Grapes of Wrath,* and *Native Son* aloud to each other, and Helen wrote on September 29, 1937: "Life much better with Scott alone."[25]

From early on in their relationship, Nearing encouraged HK to reject her bourgeois past and seek financial independence from her parents. In 1929, Nearing arranged for HK to be employed as a violin instructor at Seeds's Manumit School. Seeds disapproved of her husband's relationship with HK and soon dismissed her.

Again upon Nearing's suggestion, HK redirected her efforts to seek financial independence by getting "in touch with reality." During the winter of 1929–1930, HK rented an unheated, cold-water apartment in lower Manhattan for eleven dollars per week and worked in a series of low-wage, assembly-line jobs at a paper mill, a box factory, and a candy packing warehouse, earning thirteen to fourteen dollars per week. These jobs offered immersion experiences in the labor issues that shaped Nearing's political views. This factory work experiment also brought into stark relief the contrast between HK's newly evolving social commitments and her previously privileged life.[26]

HK returned to Europe in the summer of 1930 after her conflict with Nearing over Grace Small's pregnancy and also in adherence to his suggestion that she experience her "high life" once more to be certain that she would prefer a "low life" with him. Nearing's authoritarian directive did not rankle HK. A Maine friend and neighbor later described Nearing as "larger than life," someone from whom HK felt she could learn and whose work merited support.[27] In *Loving and Leaving the Good Life,* Helen wrote that friends of earlier years asked:

> How can you, so artistic and with such a musical and even mystical background, share your life with such a man—a pedant, a communist, an austere chap who knows so little about music and art? He will change you and dominate you.

She claimed to have responded: "To live with an older, wiser man who could answer all my questions was a continual delight; it was school and holiday all in one."[28] She also later characterized her search for a purpose or "cause":

> I felt within myself the need to follow a cause and dedicate myself to an ideal beyond the trivialities of daily suburban living . . . felt myself to be in the company of seekers. Here was a brother soul, a comrade on the way, from whom I could learn and whom I could possibly help.[29]

At age twenty-six, HK returned to Europe in order to explore the divergence between her past life and her prospective future with Nearing, which would certainly not always be a "holiday," though it might prove to be her "cause."[30]

HK traveled that summer to London, Paris, and Amsterdam, claiming that she was "welcomed and feted and proposed to," easily slipping back into "the carefree existence of the rich."[31] In a June 18, 1930, letter from Amsterdam to her parents, Helen described her arrival, "I landed in the red suit, which by the way went excellently with the red lining of [Koos van der Leuw's] car." Van der Leuw was a former suitor in Amsterdam, and he hired HK to work as his secretary at the Dutch Theosophical Society Headquarters. She reflected in the same letter to her parents, "I don't know why I've kept away from Holland so long. I love it as much and more than I ever did."[32]

Helen wrote to Scott with a similarly light-hearted tone, flirtatiously flaunting her happiness without him, "I feel a free woman. Planless as a bird, and happy as a bird, and clean and fleet-winged as a bird." He replied with a characteristically stern, pedantic critique of Knothe's notions of freedom:

1. In a world where the food and clothes you use daily are produced under the labor conditions that you sampled in NYC last spring, you cannot be free of an obligation to produce food and clothes.

2. In a world where oppression is rife, you cannot be free of an obligation to help root out slavery.

3. In a world of suffering, a person who could, and still can, relieve suffering by a touch of her hand, cannot be free of the responsibility for the use of that gift (power).

4. In a world of struggle, where members of the race are agonizing over their failures, a person who can reach people and help them to get back on their feet—a person who has a genius for reaching the heart, and refilling it with courage, cannot be free of this vital obligation.

5. In the task of uplifting the race, those in advance cannot be free of the duty to help life.[33]

Nearing criticized HK's bourgeois existence in Europe as "living in a 57th Street atmosphere," a reference to her parents' Manhattan apartment, and having a "debauch," a "fling," and indulging her "personality (lower self)." He concluded: "Did we both work so hard to free you from Ridgewood and 57th Street in order that you might turn around and walk back into Ridgewood and 57th Street?"[34] The language is reminiscent of Besant's and Lutyens's similar entreaties to distance herself from her family. Nearing insisted that HK had a "vital obligation," a "duty" to uplift the race. In so doing, she must refashion herself. HK had a propensity to respond to such a call for suppression of one's self in deference to a higher cause.

Nearing shortly thereafter sent a cable: "Enough cash granted to begin work on my book *War*. Will you come back and help?"[35] HK decided to work with Nearing, leaving lighter pursuits for demanding work and Nearing's extreme austerity. Upon her return to New York, they moved together into a three-room, cold-water walk-up flat on the fifth floor of a tenement building at Avenue C and Fourteenth Street in Greenwich Village. The toilet was in a chilly hallway and the small bathtub was next to the kitchen sink. HK described the nature of their relationship:

> Our intimacy widened easily and naturally until we were lovers as well as friends, yet sex never played a predominant part in our togetherness. Our main feeling was compatibility in thought and action—in trust, consideration and respect.[36]

She also described the ease with which they moved in together, despite social prohibitions against unmarried couples doing so. Such rebelliousness and unconventionality were characteristic of HK and reminiscent of the independence that she had displayed in her travels with Krishnamurti.

HK worked hard as Nearing's secretary, coordinating finances, editing manuscripts, mailing pamphlets, and scheduling his speaking engagements. Nearing wrote the following playful request in December 1931, illustrative of their working relationship:

> Sweetheart,
>
> I need a helper—May 15–September 15 or October 1, 1932. Someone who can type, file, etc., and is moderately sweet tempered. Not a secretary, but a co-worker. Salary $100 per month.
>
> The job has to do primarily with a manuscript called "Worldism" on which I am working. Der Germany or Russia.
>
> Do you know anyone who would be interested or who could be induced to take this on? If so, please let me know.
>
> You are a darling, Scott[37]

Nearing paid HK a wage for her secretarial work that earned her some semblance of economic independence. Yet HK remained for the most part financially and emotionally dependent upon Nearing—her employer, mentor, and lover—which was arguably a more precarious reliance than upon her parents. HK and Nearing wrote faithfully to each other, and Nearing expressed concerns regarding his potential dominance: "You are a vigorous, well-developed person who needs room to grow in. When I am around you, you do not always have room enough." He signed this letter, "I love you. You are my dear and close friend and comrade," denoting recognition of his power as well as deep affection and implied partnership.[38]

HK and Nearing shared not only a romantic and professional bond, but also a spiritual connection. In their initial telephone conversation in 1928, Nearing inquired at length about HK's theosophical studies. With a keen religious sensibility rooted in practical Christianity, Nearing believed firmly in deepening and strengthening the spirituality of a citizenry.[39] While at the University of Pennsylvania, Nearing had been an active member of a Baptist church and briefly considered a vocation in the ministry, though he ultimately left organized religion, concluding that the church could not respond adequately to the social crises of the day.[40]

Nearing had participated in a number of séances in Toledo, claiming to have communicated with several of his personal heroes, including the Russian novelist Leo Tolstoy. He was intrigued by HK's early, rigorous occult training at the Manor and her close relationship with Krishnamurti. HK encouraged Nearing to meditate, and she became his guide in this area.[41] In turn, Nearing strongly supported HK in honing her spiritual skills, writing to her: "With your reading and study in the occult, you are in a position to make a very important contribution just now, when this field is so sadly neglected and is in such very great need of attention." Nearing suggested that HK develop her public persona as well, claiming, "You have important work to do, and I think speaking and writing will be an essential part of it."[42] He further encouraged her to spend time each year in Holland if her "etheric body is at one with the atmosphere" there.[43] They shared a deep commitment to spiritual growth, and Nearing designated HK the spiritual leader in their work together.

Nearing, however, would be the political instructor. HK later wrote, "Politically and practically he was the leader; in the more intangible artistic and spiritual domain, I was the guide."[44] She described herself as a "political ignoramus" with her head in the clouds, diminishing her penchant for astrology, palmistry, and graphology as less weighty compared to Nearing's concerns for imperialism, social justice, and war. HK persistently cast herself as below or secondary to Nearing, particularly in the realm of politics. Such a self-deprecatory posture was again reminiscent of her deference to Theosophical Society Masters.

In the summers of 1931 and 1932, HK received her first overseas political lessons from Nearing when they traveled to Germany and Russia. Helen's letters to her family from Berlin and Leningrad were politically charged as she invoked Marxist rhetoric, espousing Nearing's economic and social views and spouting radical political theory with a youthful, student-like zeal.

At that time, Berlin was at the epicenter of escalating political tensions in Europe. Hitler's National Socialist Party's rise to power and fascism's rapid spread were threatening the peace accorded by the Treaty of Versailles at the end of World War I. HK characterized Berlin as "the storm center of the new dictatorship" and wrote to her parents in August 1931 describing the suppression of the communists whom HK and Nearing were there to support.[45] She depicted one woman as having been assaulted by police for distributing communist propaganda:

> I heard a stampede outside my window, jumped up and saw two policemen chase and catch a woman, beat her with clubs, knock her about brutally and march her off to jail for distributing Communist leaflets.

HK described Nearing's response as "cautious," but claimed zealously, "I want to get in the middle of every street fight I see."[46]

From Leningrad, HK wrote to her parents in praise of the Russian workers whom she saw thriving under communism:

> These people are so fresh, clean, vigorous and enthusiastic, entirely selfless in their work. Of course, one finds such also in Western lands, but never to such a degree or so many. And where could workers in America avail themselves of every resource of such a finely equipped place? And who but the workers should have such advantages put at their disposal? Surely not the debutante lap dogs nor pining society women—yet such it is at home.

HK disparaged "stuffed bourgeois minds," claiming "You won't find me pleasant company when I come home" as her embrasure of Nearing's communist radicalism meant a rejection of her Ridgewood life.[47] HK even suggested to her parents that they need not pick up her voter's registration in Ridgewood as Nearing had told her that they "throw out communist votes."[48]

HK embraced Nearing's politics like a passionate apprentice. She wrote a similarly charged letter to a former Theosophical Society friend and suitor from Germany in the wake of her European travels. He ridiculed HK's newfound zeal for communism, responding "Don't try to kid yourself. You have found a man with extreme charm, intelligence and strength, and you have made his mania your own." He added, "We aren't all the nasty bourgeois frumps, as you seem to think, and you aren't the robot you imagine yourself to be, thank goodness."[49] HK's attraction to Nearing's oppositional stance was genuine, but her political fervor in many ways seemed merely an exaggerated expression of Nearing's philosophies.

Though HK's political rhetoric may have been affected, the strenuous work that she took on in their relationship was authentic and her own. Between 1932 and 1940, HK and Nearing began developing a rural property in Vermont that required ongoing labor and attention. Furthermore, HK's secretarial workload increased steadily, particularly through the 1940s. Nearing issued a monthly news commentary from 1942 to 1949 entitled *World Events,* and he also published numerous social science handbooks: *United World, The Soviet Union as World Power, Democracy Is Not Enough, The Tragedy of Empire, War or Peace?* and *The Revolution of our Time.* HK and Nearing's letters to each other at this time continued to denote affection and mutual respect, though they teetered between love notes and business correspondences, demonstrating a balancing act that they would sustain for fifty-odd years.

HK took on Nearing's daily regime and arduous work habits. In a letter to her parents, she related their schedule during a retreat in Ehrwald:

> We're up at six, out and ski for an hour and watch the dawn break, breakfast on raw oatmeal soaked over night in milk (from the cow downstairs) and soaked figs, prunes or apricots. We work from nine to two and then down to the village to shop, back and eat a raw salad and bread and butter and honey, out and ski some more, chop some wood until it's dark, and then read by the fire till nine or so.

This balance between work and leisure—with emphasis upon healthy food and physical activity—epitomized Nearing's personal discipline. HK readily adopted his routine because she believed their days were spent purposefully and productively. She wrote to her parents, "A simple life but just exactly what we both want and like."[50]

Their primarily vegan diet, often eaten with wooden chopsticks and out of wooden bowls, became central to their life together. HK proved particularly vehement in her condemnation of meat-eaters, though she frequently diverged from veganism with her pronounced weakness for ice cream.[51] HK and Nearing in their later writings touted their organic diet, deeming processed foods such as bleached flour, white sugar, and polished rice "poisons" and eschewing "habit-forming drugs" such as caffeine, cola nut extract, nicotine, and alcohol.[52] They ate and served guests large bowls of raw fruits and vegetables, grains or simple soups. Some days they followed a "mono-diet" in which they would eat only one item such as apples throughout the course of a day. HK and Nearing also fasted regularly or drank a liquid, fruit juice diet in order to rest their digestive systems.[53]

While traveling, they would try to maintain their diet by eating in their hotel rooms rather than at restaurants or banquets. Nearing wrote to HK from the Washington Hotel in Chicago:

> I have eaten two meals here today (in my room) and shall eat a third in a few minutes. Oranges for breakfast. For lunch ripe olives, nut cheese, whole wheat bread, honey and apples. Tonight I shall have some oranges and dates. (also I ate some lettuce for lunch).[54]

They practiced "wellness" based upon this vegetarian diet in combination with physical fitness and productive professional work.[55]

However, HK tempered Nearing's strict regimen, bringing out what she referred to as "an unpredictable, light, and perceptive side of his nature." For example, she described skiing together during their travels in the Austrian Tirol. Upon HK's suggestion, they skied nude and "sped down the slopes wearing only big ski boots and skis."[56] HK infused Nearing's rigidity with wit and whimsy.

Helen Nearing's political formation under the tutelage of Scott Nearing propelled her development as a rebellious, spiritual woman. Nearing encouraged her to realize her "purpose" through their austerity and hard work. Health, vigor, and simple living shaped their daily regime and would prove central to their homesteading life in Vermont, where they would begin to create their good life story.

CHAPTER 3

The Vermont Experiment

Forest Farms and Maple Sugar ✤

BEGINNING IN THE EARLY 1930S, Helen Knothe had grown increasingly disgruntled with her life in New York City, disparaging both the city and their dank, walk-up apartment on Avenue C as "dreadful." Many of HK's days in the city were spent doing research or "librarying" for Nearing at the New York Public Library, typing Nearing's manuscripts, playing the violin, or shopping with her sister Alice. She attended Theosophical Society meetings, concerts, movies, and periodic parties, the latter of which she characterized as a waste of time.[1] Though days were full, HK struggled to find a clear sense of purpose in her urban existence.

Nearing likewise was discontent with life in New York. In 1931, he terminated all formal ties with the organized left, resigning from executive positions in the Workers' School, the Garland Fund, and the Civil Liberties Union. Attendance at his lectures was in decline, and he was estranged from American communist thinkers and writers, drawing staunch criticisms such as "Mr. Nearing exaggerates the simplicity of the solution he has to offer . . . He is wrong in insisting that Russia offers us a direct and obvious alternative."[2] At odds with main-

stream politics and extreme in his pro-Soviet stance, Nearing concluded that individual liberation could only be realized apart from
capitalist society. He ended all institutional affiliations in his search
for personal freedom.[3]

HK and Nearing thus began to gravitate toward a rural alternative.
They looked for a cooperative or intentional community that they
might join in Vermont, eventually purchasing the Ellonen Farm in
the Pikes Falls region of southern Vermont on December 6, 1932.
Their turn to a natural landscape was emblematic of a larger social
movement at the time. Rooted in Jeffersonian ideals and nineteenth-
century Western settlement, the practice of homesteading continued
in the United States throughout the twentieth century, with the number of homesteaders surging during times of economic crisis such as
the Great Depression. Homesteading offered a means of social protest
against the culture that produced the economic predicament.[4] The
Progressive Era and the Gilded Age had seen an American revival of
interest in nature as something to be visited, experienced, and pos-
sessed.[5] During the Great Depression, however, the natural world and
its concomitant homestead became a "panacea" for troubled times.[6]
Pastoralism, idealized in literature such as Pearl Buck's *The Good
Earth* (1931), prompted a second wave of conservation politics steeped
in "ideology of the folk" and in notions of a self-sufficient common-
wealth at work on a farm. Nature and this rural aesthetic offered spiritual renewal, as well as home-based moral and social reform. The city
was the antithesis of this idyllic world; nature was sacred, a "true set
of laws" around which daily life could be ordered.[7]

HK and Nearing's gradual shift to a rural setting was thus part of a
larger cultural turn to the land. Their retreat to the country was also
an expression of pacifism in light of larger global crises. HK described
their early homesteading experience in her 1936 to 1940 diary against
the backdrop of intensifying world conflicts. She accurately depicted
Hitler's 1936 entry into the Rhineland as a precursor to war; fascism
was ushering in a "Dark Age." HK described having nightmares about

being one of Hitler's chosen followers in the summer of 1936. She referred in her diary to Hitler's marching into Austria in 1938, to the German invasion of Denmark and Norway in April 1940, and to her inability to sleep after Hitler invaded Holland in May 1940. Her May 17, 1940, entry spoke of her profound shift to a rural landscape in the midst of such turmoil: "Nazis arrive Belgium, planted sweet peas."[8]

Scott Nearing previously had chosen to seek refuge close to the land. In the early 1920s, he and Seeds moved their family from New York City to the Nearing farm in Ridgewood, New Jersey. He converted the barn on the property into a house, adding a large stone fireplace as well as an organic garden and woodshed in the back.[9] He also had spent the summers from 1906 through 1915 as part of the Arden community, a Delaware-based, single-tax rural community that challenged the dominant values of industrial capitalism through simple, cooperative living. Nearing described "bread labor in contact with nature," work tied to land, as the most rewarding aspect of Arden, where he built his first stone structures and cultivated an organic garden. He called his stone house at Arden "Forest Lodge" and later described the experience as the "good life in miniature."[10] Upton Sinclair lived in a tent next to the Nearings in the summers and rented their house when they returned to Philadelphia during the winter months.[11]

HK's move to rural Vermont marked a departure from her primarily urban and suburban existence. Their modest, drafty clapboard farmhouse was far removed from her Cottage Place home. HK's sister, Alice Knothe Vaughn—a "country clubber" who preferred to play bridge and golf—described Helen's rural lifestyle as "weird." Vaughn claimed to have been afraid of Nearing and not to have understood what her sister saw in him.[12] Vaughn nevertheless sent her daughter Cornelia to spend several weeks each summer with HK in Vermont. Cornelia's father, Charles Vaughn, would burn her chopsticks upon her return in condemnation of Knothe's and Nearing's excessive radicalism. However, Cornelia reflected back on her visits as a "happy time," and the Vaughns moved to Manchester, Vermont, in the mid-forties. They were attracted both by the beauty of the state and by Knothe and Nearing's example of a simpler existence, though they would never be as extreme or austere. Helen's brother Alex likewise moved to Vermont in the late forties, demonstrating a tendency, first evidenced by the Knothes' original move to Ridgewood, for the Knothe family to be drawn to more natural settings.[13]

HK quickly came to prefer daily life in Vermont to the "hurly burly" city. She wrote in her diary on January 10, 1936, that she was ready at age thirty-one to retire there. HK relished the woods and the cross-country skiing, skating, swimming and sunbathing, despite bouts with lice and many harrowing, through-the-night drives over the hills into Vermont. "This is the life," she wrote on May 10, 1939.

HK contributed her spiritual expertise to their back-to-the-land endeavor. Rebecca Kneale Gould described homesteading as almost a liturgical discipline, an "extra-ecclesial religious practice" and "dissent from mainstream culture" imbued with spiritual meaning. According to Gould, Knothe and Nearing's farms were a "seminary in the woods" in which they were "first acolytes and then spiritual leaders."[14] Helen—the dowser and mystic—acted as purveyor of the occult on their homestead.

HK also contributed financially to their land acquisitions. She was able to purchase her shares through an inheritance of 5,000 guldens (approximately $35,000 at the time) left to her by Theosophical Society friend and suitor Koos van der Leuw, heir to a chocolate concern in Holland who recently had died in an airplane crash.[15] Nearing claimed to have cashed in an old life insurance policy to pay his part. HK and Nearing purchased the Ellonen Farm in Nearing's name, a property which included seventy-five acres in Winhall, Vermont, and ten acres in Stratton. Nearing put down a $300 deposit and assumed an $800 existing mortgage, which was paid in full by December 28, 1933. HK and Nearing expanded their property significantly, purchasing over nine hundred acres of land in three neighboring towns between 1932 and 1935—in the heart of the Great Depression. On January 2, 1933, Helen Knothe purchased fifty-seven acres from Jonathon Tibbets in Jamaica, and Nearing purchased two adjoining parcels (four hundred and fifty acres) the same day, both without a mortgage. On December 12, 1933, Scott Nearing purchased two parcels in Winhall (215 acres)—again without a mortgage—from the Linscotts. On December 5, 1934, HK signed a sales agreement with Mercy Hoard for seventy-five acres in Winhall at $1,600, which she paid off in full within the week. Nearing purchased thirty-five acres more from the Hoards on April 12, 1935, and previously had purchased a thirteen-acre sandpit from Sarah Clayton for $135 on August 25, 1933. In sum, over the course of three years, the couple spent more than $8,000 for 930 acres in this rural Vermont region at a time when a day laborer might earn $600 for a year's work. In 1937, they began selling property, and by the end of World War II, had sold off seventy-three of their acres for $5,700, recovering two-thirds of their original land cost but still owning 92 percent of the land that they had purchased.[16] They also maintained their Avenue C apartment in New York City until 1940, at which time they became year-round residents in Vermont—though Nearing often traveled as many as four of the winter months.[17]

These land purchases reveal the extent to which HK and Nearing brought independent capital to their homesteading project. Beyond her inheritance from van der Leuw, Knothe probably received some portion of her family's profits from the sale of Knothe Brothers Company in the early 1930s. Through savvy land acquisitions, her inheritances, and profitable maple sugar seasons, HK began to live out Nearing's edict for financial independence. He wrote to her on February 4, 1940:

> Incidentally, in 1928–9, you were economically dependent. Now you have your own place and if you are willing to work 4–6 months per year, a modest income sufficient to cover the needs of the entire year. This gives you an economic base—quite free of any debt—with low overhead costs.[18]

HK purchased significantly fewer acres—132 acres to Nearing's 798—yet she owned the Hoard property upon which they would build their primary residence and which held the most value.

HK and Nearing formulated a ten-year plan for their property they originally deemed "Forest Farms," though by 1950 the name would evolve into "Forest Farm." They viewed their valley in Vermont as a laboratory for "testing out certain principles and procedures" derived from Nearing's economic and social theories. They would avoid debt, barter when they could, and seek cooperative ventures. Through self-discipline and clearly defined work schedules—four hours a day devoted to bread labor or basic work, four hours to professional interests, and four hours to responsibilities as citizens—HK and Nearing would cultivate an organic garden to support their vegetarian diet. They anticipated the broader organic gardening movement that would burgeon in the United States during the 1940s.[19]

HK and Nearing used simple, second-hand tools, which Nearing meticulously maintained, and they avoided the use of domesticated animals, though they initially relied upon horses for labor. They sought a "no-money economy" and, with clear budgeting and planning, avoided glaring inequities in work status, involving all members of the household in service as exchange for income. Theirs was to be an extraordinarily austere, frugal, out-Yankeeing-the-Yankee existence as they would adhere to Nearing's mantra: "Live within your income; spend less than you get; pay as you go."[20]

Despite their goal of a "no-money economy," HK and Nearing recognized the need for a cash crop in order to make their Vermont life self-supporting. The Hoard farm that they purchased in 1934 had an established maple sugar bush. The property also had a decrepit sugar house that continued to be run by Floyd and Zoe Hurd with their thirteen children "working bush on shares"—exchanging labor and equipment for a portion of the eventual harvest. HK and Nearing worked cooperatively with the Hurds for the next six years, ironically contributing to the employment of their sizable child labor force.

They maintained two thousand to twenty-five hundred taps and used a horse-drawn wagon to transport sap buckets as well as the fifteen to sixteen cords of wood required annually to fuel the evaporator. After a promising maple sugar crop in the spring of 1934, HK and Nearing purchased another adjacent tract of land that significantly augmented their sugar bush. With the help of Vernon Slasson, Nearing organized the clearing of the maple grove, engineered seven miles of galvanized pipes to carry sap down from the woods, and coordinated the construction of a new sugar house in 1935.[21]

HK contributed to all areas of the project but, in particular, oversaw the boiling down of the sap and the pouring of syrup and candy, enlisting a necessary work-crew for both endeavors. She also assumed responsibility for the sale and distribution of their maple sugar products, bringing her whimsy and flair to the enterprise. HK selected quaint syrup jars shaped like Dutch women and capitalized on family connections by enticing a handful of Fifth Avenue shopkeepers to sell their goods.[22] One Vermonter described Scott Nearing as "the best damn capitalist I ever met," and HK, with her marketing skills, earned a similar reputation.[23]

HK's workload in Vermont was varied and formidable as she developed a number of skills beyond sugaring and marketing. HK assumed the "inside work" of cooking, canning, and cleaning, while her secretarial duties and household management responsibilities expanded as well. HK and Nearing attempted to simplify indoor jobs by wearing second-hand clothes that did not warrant repair, by minimizing house cleaning, and by eating a basic, raw-food diet.[24] HK also paid a neighbor to do their laundry under the auspices of making a contribution to a poorer family's income.[25] Nevertheless, her load was substantial, with 1939 marking a particularly demanding sugar year during which she enlisted extra help from her parents with the pouring and packaging of syrup and candy. HK described the work as "hard" but "fun to do."[26]

HK and Nearing still sought to balance their work and health. Both touted the benefits of Vermont's fresh air, and HK continued to espouse the healing benefits of solar rays. They frequently sunbathed nude, being "one with the forces of light in the universe."[27] HK later described her adaptation to the country with a characteristically romantic, favorable gloss:

> To live in the country was something new for me. Summer vacations on lordly estates or at summer camps beside lakes were more my custom, but I adapted well to the enforced frugal life . . . I adapted surprisingly well to Vermont backwoods conditions. I even came to relish privations.[28]

However, in truth, her early adjustment to life in Vermont was checkered with illness.

During the winter of 1934–1935, HK wrote to Nearing describing painful hemorrhoids and bouts of depression.[29] Over the next several years, she continued to struggle with symptoms of pain and fatigue reminiscent of those she experienced in Vienna in 1923. Isolation and anxiety may have been contributing factors as the symptoms occurred mostly during the winter months while Nearing was traveling. Lack of light also may have precipitated her depression, as HK's healing rays would have been eclipsed by short winter days.

In 1935, HK underwent a physical examination by the well-known health specialist Dr. Philip Lovell at his Health House in Los Angeles. She wrote to her mother:

> He found me in fine condition. Nothing wrong—except that he doesn't think I can have a child. Found the left ovary cystic and thinks the chances are twenty to one I'm not fertile. Too bad for my purposes, but there it is. I'll talk it over with Scott.[30]

HK's resigned tone in her letter to her mother masked enormous disappointment as she had appealed persistently to Nearing for a child and would continue to do so despite Dr. Lovell's prognosis. In a 1940 letter to Scott, Helen wrote, "I've asked you time and again to give me a child. That would settle me down. But I know the difficulties. I'm only reminding you." He responded by suggesting that she find someone else to marry and have a child with, if that was what she wished.[31]

Nearing also continued to have sexual affairs with other women, which increasingly disturbed HK, as evidenced by a January 1944 letter:

> Scott, oh Scott, even into this nice letter of warm thoughts comes the ache and wormlike thought: The warm unity and closeness is gone. Someone else also is writing you thus and thinking thus and wanting to be close. And I no longer have any safe feel or right to ask or feel myself close. See how I've spoiled even this letter now. But I'll let it go off. You might as well know the depths of my pain and how it hangs on. The complete break with Krishna never hurt like the crack in our relationship . . . Your Good Friend, if nothing else.[32]

Nearing responded with a subtle critique of HK's being capable of jealousy and his surprise regarding her feelings of insecurity. Their emotional exchange revealed the sexual tensions in their relationship that had persisted since 1929, for Grace Small was merely a precursor to future liaisons.[33] HK later wrote, "People say it's a shame I never had children, but I believe in reincarnation and have lots of children in other lives, so I don't feel unfulfilled." However, in these early years, not having a child, not being married to Nearing, and not having a faithful partner were sources of great pain for her.[34]

The 1940s proved particularly formative for HK as she confronted these complex issues, often seeking spiritual solutions. In January 1944, HK underwent surgery related to her ovarian cysts detected by Lovell.[35] Shortly thereafter, she experienced her first "automatic writing"—an occult experience in which HK became a channel for spiritual messages that she then transcribed. HK later described the ensuing period of repeated automatic writings as an "excellent time," a spiritual watershed during which she communicated with Theosophical Society characters Master Morya, the deceased Nitya, and Koos van der Leuw. These old associates told her that, despite Nearing's protestations, she and Nearing would be the parents of the reincarnated van der Leuw or Nitya or Helen's dying father, and their child would be a girl.[36]

At the end of April 1944, HK flew to Orlando, Florida, to be with her father during his final days. She believed that she was then pregnant with Nearing's child, as her automatic writings had foretold.

Helen Knothe with her brother Alexander Knothe, mother Maria Knothe, and new sister Alice Knothe (1907).
THOREAU INSTITUTE.

Helen Knothe with her father, Frank Knothe (1922).
THOREAU INSTITUTE.

Radha Rajagopal Sloss, Helen Knothe, Krishnamurti, and his brother Nitya (ca. 1923). THOREAU INSTITUTE.

Helen Knothe as a Theosophical Society pupil (ca. 1923).
THOREAU INSTITUTE.

Helen Knothe and Theosophical Society friends in Ehrwald, Austria (1923).
THOREAU INSTITUTE.

*Helen Knothe (far left) beside the lotus pool at the Theosophical Society
headquarters in Adyar, India (1924).* THOREAU INSTITUTE.

*Helen Knothe hiking in
Australia (1925).*
THOREAU INSTITUTE.

Scott Nearing (ca. 1928).
THOREAU INSTITUTE.

Helen Knothe with Masha Nearing,
John Nearing's wife, in Russia (1931).
THOREAU INSTITUTE.

Forest Farms, Vermont (ca. 1950). THOREAU INSTITUTE.

*Helen Knothe Nearing singing at a Sunday morning
gathering in Vermont (ca. 1950).* PHOTOGRAPH BY
REBECCA LEPKOFF. THOREAU INSTITUTE.

Helen Knothe Nearing pouring sap (ca. 1950).
PHOTOGRAPH COURTESY OF RICHARD GARRETT.

Helen Knothe Nearing in a promotional photograph for The Maple Sugar Book *(1950).* PHOTOGRAPH BY LLEWELLYN RANSOM. THOREAU INSTITUTE.

Scott Nearing, Helen Knothe Nearing, and a friend in India (1967).
THOREAU INSTITUTE.

Above: In their first Maine Forest Farm house (1976).
PHOTOGRAPH BY RICHARD GARRETT.

Left: Helen Knothe Nearing and blueberries (1970).
PHOTOGRAPH BY RICHARD GARRETT.

Above: Building in Maine with the Ernest Flagg slip-form method (1970). PHOTOGRAPH BY RICHARD GARRETT. THOREAU INSTITUTE.

Right: The Nearings' wooden bowls and spoons. © PHOTO BY LYNN KARLIN.

The kitchen in their final Forest Farm. © PHOTO BY LYNN KARLIN.

Helen Knothe Nearing beside her yurt with Juliette Baïracli-Levy (1990).
PHOTOGRAPY BY RICHARD GARRETT.

The greenhouse and compost piles at their final Forest Farm.
© PHOTO BY LYNN KARLIN.

Helen Knothe Nearing and a favorite stone (ca. 1990). THOREAU INSTITUTE.

Helen Knothe Nearing typing at her picture window, overlooking Spirit Cove (ca. 1980). THOREAU INSTITUTE.

Helen Knothe Nearing and a favorite poppy flower (1989).
© PHOTO BY LYNN KARLIN.

However, HK described the aftermath of Frank Knothe's death on May 4, 1944, in her Spanish Notebook years later:

> Down to Fla. F died.
> Back to Maine [Vermont]; all gone.
> Bad time. Girls.
> Bldg, sugaring, selling, tapping.
> Visitors, Visitors.

The spiritually charged, excellent time of automatic writings was "all gone," as was her supposed pregnancy. "Girls" likely alluded to Nearing's continued infidelities, and her list of homesteading responsibilities along with "visitors, visitors" indicated that it was a period of overwhelming work.[37]

Yet much of the homesteading work proved genuinely rewarding for HK, particularly building with stone. Nearing had written to HK in 1931 describing their life together as a metaphorical building project:

> This is our life, Sweetheart. We are out of the kindergarten. We have taken our facts on our backs, our tools in our hands. You and I have work to do. A structure to build. Let's build, day by day, my own. And as we lay a stone, let's be sure that it is in place, it fits well. Let's make a superbly good job of it, Helen.[38]

In Vermont, they literally built structures together—thirteen buildings in all—out of native stone and wood using the Ernest Flagg slip-form method. Ernest Flagg was a New York architect who in his 1922 book *Small Houses* described a low-cost method for stone work using stone and movable forms that allowed "people of limited means and experience" to build permanent dwellings. Knothe and Nearing later wrote that it was their intention to use this Flagg method to create an organic architecture in which their buildings would be adapted fully to their environment as form and function united.[39]

They initially constructed a stone retreat for Nearing in the woods behind the Ellonen farm house and also added a stone wing and fireplace to their clapboard farmhouse. HK assumed the tedious work of pointing or finishing with mortar. After the 1935 purchase of the

Hoard Farm, HK and Nearing began constructing several stone buildings on that site. They first built a lumber shed onto which they would later add a two-car garage and storage area. They next built a guesthouse on the foundation of the razed Hoard farmhouse and added a woodshed with a stone and concrete wall on three sides. Finally, between 1939 and 1942, HK and Nearing constructed their main house, incorporating a large boulder as the rear wall, the exterior portion of which served as an outdoor deck for gatherings. They added a stone cabin for their friend Richard Gregg, a stone and concrete tool shed with a sun-heated greenhouse, and a stone outhouse, using the same forms for all buildings.[40] These projects were pivotal first building experiences for HK. She would boast later: "I handled most of the stone," while Nearing mixed the concrete, built window and doorframes, hewed beams, and roofed the buildings.[41]

HK exaggerated the extent to which she and Nearing were self-sufficient builders as they hired a number of co-workers from the community. Delbert Capen, for example, performed many duties such as hewing beams and raising concrete walls, while George Wendland, a German master craftsmen, contributed his building expertise.[42] Denial of the help that they received denoted a growing idealization and mythologizing of their homesteading life, particularly on the part of HK. It also was rooted in a general antagonism toward local folks in the Pikes Falls Valley. In a later description of their Vermont community, HK and Nearing criticized the area as being "typical of rural America where separatism and individualism have subdivided communities almost to the point of sterility."[43] Their disdain for the local people was reciprocated as Vermonters distrusted Nearing's communism and radicalism. HK and Nearing were outsiders and had become extensive landowners. Vermonters, with their rugged independence, "had no use for you if they didn't know your grandfather" and would have been wary of such rapid land acquisition and evident wealth.[44]

HK and Nearing furthermore were perceived in the community as excessively stern taskmasters.[45] A neighbor who helped HK package maple sugar candy described her as an imperious overseer: "Her temper was so quick, so sharp. And NO one put the wrong sugar candy into the wrong color cellophane. It didn't matter if you were eight or ninety, you were set right."[46] Another neighbor noted, "There was a coldness to Helen, there was an officiousness about her," that generally distanced her from the local community.[47]

Nevertheless, HK and Nearing were generous with children in the area, bringing them Christmas presents and hosting Easter egg hunts.[48] Moreover, in a few instances, HK and Nearing described experiencing genuine cooperation with their Vermont neighbors. For instance, HK worked with Pikes Falls residents to recover their discontinued postal delivery in 1945. Fourteen households, including Forest Farms, had lost postal delivery because of war-time cutbacks. HK was appointed secretary and Nearing was chosen as one of three advisors to the Pikes Falls' Citizens Committee charged with waging a protest against the discontinuance of delivery. HK masterfully publicized the community's plight as a David and Goliath story, a rural town cut off from mail service and unable to correspond with their boys and girls in the service. HK and Nearing also were dependent upon mail service in order to promote their maple sugar products, to transport Nearing's manuscripts, and to communicate with each other during their respective travels.

Community members wrote to their congressional representative, and both the *Boston Globe* and the *New York Times,* along with local Vermont papers, ran articles about the community's plight. HK posed for a *Boston Globe* photograph in which she carried baskets of mail. The caption read, "Helen Knothe, heavily laden with mail, starting out from post office over eight-mile route to Pike's Falls"—an exaggerated display on her part but an effective tactic. Within two weeks, they had regained service. HK and Nearing described this as the only time during their twenty years in Vermont that the community demonstrated common will and cooperation.[49] However, the cooperation remained under the direction and control of HK and Nearing and was reliant upon their marketing expertise.

HK interacted more frequently with the like-minded pacifist and artistic subculture that was developing in the Pikes Falls Valley during the thirties and forties. Comprised mostly of urbanites from New York, Pennsylvania, and New Jersey who came as both summer and year-round residents, this alternative community was attracted to the area by established residents such as Scott Nearing, author Pearl Buck, and environmentalist Richard Gregg.[50] Harold Field of the Dupont family and his wife bought an old farmhouse and ninety acres in the area in 1941, leaving Philadelphia to try to live on the land. Photographers Rebecca Lepkoff and Irene Strauss, active members of the New York City Photo League, built summer homes with their families.[51] Norm Williams, from a missionary family that had lived in an ashram with Gandhi in India, purchased with his wife Winnie a remote property on which they built a stone house named *Wilken Croft* ("cloud scratcher"), constructed wholly of stone and wood from their land. Williams, an extremely civic-minded accordionist, donated land for a community center for folk dancing and local gatherings.[52] HK and Nearing had sponsored Hilda and George Wendland as refugees from Germany during World War II, and they became part of this subculture. Hilda Wendland grew to be a particularly close friend of Helen's because they shared a Dutch heritage as well as mystical and musical interests.

Hospitality toward this artistic, pacifist community became central to HK's homesteading experience, despite her penchant for solitude. She and Nearing hosted Sunday morning music gatherings, for which HK provided music books and played the recorder or violin. They also held Monday evening lectures at Forest Farms, during which Nearing directed political discussions or lectured on current events.[53]

HK and Nearing further provided the "free hotel" in town. Espousing hospitality as a great teacher, they provided meals and a place to stay to anyone who needed them, following the Dorothy Day model. Day was a contemporary of HK and Nearing's who had spent her youth among anarchists, socialists, and communists and her adulthood as a writer and radical. Day's radicalism grew out of a deep sense of injustice suffered by the poor; she was moved by Christian

Gospel stories in the wake of a personal conversion to Roman Catholicism. She saw in Christianity a possibility for freedom, community, and solidarity beyond the reach of politics. In turn, Day assumed a life of voluntary poverty marked by having nothing one fears to lose, by reducing self-interest, and by finding security outside of material values. She founded the periodical *The Catholic Worker*, which she first distributed at a Communist rally in Union Square in New York City on May 1, 1933, and to which HK and Nearing subscribed. Day urged the establishment of Houses of Hospitality in every Catholic parish to care for the homeless and unemployed. The Houses of Hospitality were modeled after European hospices or free guest houses and burgeoned in the same way that bread lines did in the height of the Great Depression as word spread that there was always coffee, soup on the stove, and all were welcome.[54]

HK and Nearing emulated this hospitality model in Vermont, though they enlisted work from their guests in return. The visitors performed odd jobs ranging from peeling apples to chopping wood to building houses. Sculptor Jerry Goldman was a former guest who noted Helen's remarkable organization and skills as a host, illustrated by her having a central table on wheels that she could shift out of the way with her hip as she moved quickly and efficiently around the kitchen. He also decribed Knothe and Nearing's primitive toilet system in their main house, which required a user to fill two buckets from the hand pump in the kitchen and carry the buckets through the living room into the bathroom. Goldman characterized this ritual as having been purely affected, an overt demonstration or performance of their simple living. HK and Nearing could merely turn on a faucet to water their garden.[55]

December 1946 marked a notable shift in HK's relationship with Nearing as Nellie Seeds Nearing passed away. Scott Nearing visited Seeds the day before she died. Their son Robert claimed that she told him, "Oh, Scott, I'm putting off all evils from along the way. I forgive you for Helen."[56] The next year, forty-three-year-old Knothe and sixty-four-year-old Nearing were married. She later wrote in *Loving and Leaving the Good Life* that they happened to be in California where Nearing was giving a series of lectures when they decided to marry. She claimed that as they walked to the town clerk's office, Nearing asked, "Do you want a ring?" to which she responded, "No." "Do you want flowers?" "No." And afterwards they rarely used the words "husband" or "wife."[57] Yet in a December 14, 1947, letter to her mother, Helen sounded more sentimental:

> Of course I'm having a wonderful time and of course I'm happy as a bird. It *is* a relief to have the ceremony performed and I feel more re-laxed and secure. Though I was willing to go on as I had for years. Scott is the man I love and admire most in this world and it's nice to be linked with him publicly. If he gets in trouble, and he may soon, I want to be with him and now I can.[58]

Helen Knothe Nearing [hereafter HKN] indicated both a desire to share in Nearing's radicalism as well as a prior insecurity in not having been his wife.

HKN spent the following winter alone in Vermont writing *The Maple Sugar Book*. The book was a detailed, autobiographical account of their work based upon HKN's years of research on maple sugaring. Their Vermont neighbor Pearl Buck had encouraged HKN to write about their maple sugar project, and Buck's husband Richard J. Walsh, President of the John Day Company, agreed to publish the book in 1950. It was printed as a co-authored text but was "primarily Helen's handiwork" and her first foray into autobiography. HKN later claimed that without Pearl Buck's encouragement, she might not have written the text because "we were busy living it."[59] HKN main-tained that the purpose of the book was threefold:

> We had three things in mind when we set ourselves to write this book. The first was to describe in detail the process of maple sugaring. The second was to present some interesting aspects of maple history. The third was to relate our experiment in homesteading and making a liv-ing from maple to the larger problem faced by so many people nowa-days: how should one live?[60]

The text diverged from Nearing's politically charged tracts and represented their initial prescriptive guide to purposeful living.

The Maple Sugar Book was described as "the first complete treatise on the maple industry."[61] Its publication established the Nearings—formerly "city folk"—as spokespeople for homesteading and maple sugaring, bringing HKN into the foreground as the writer, promoter, and marketer of their story.[62] Rebecca Kneale Gould described the Nearings' book as a traditional conversion narrative with nature becoming their sacred world, and consumerist, capitalist society the sinful one. For HKN, the narrative itself was her conversion and rebirth, as she became author, actor, and publicist of their story, recasting their homesteading life as a public endeavor.[63] She found her purpose in the performance of their voluntary, self-sufficient living and would stand on her own as a superlative marketer of their homesteading ethic.

John Day Publishers held a "sugaring off" party on February 20, 1950, at its Manhattan office, to which the Nearings brought thermoses of Vermont snow for a traditional sugar-on-snow dessert. HKN, in promotion of *The Maple Sugar Book,* gave her first extended speech in Barre, Vermont, at the Vermont Sugar Makers Association annual meeting.[64] She offered public readings, and both Nearings participated in radio talk shows, though HKN stood out as having "a wonderful radio personality."[65] Book reviews ran in the *Boston Sunday Globe,* with a photograph of the Nearings pouring sap and in the *New York Times,* describing the "distaff side" of the couple spending the morning in the *Times's* kitchen offering samples. Although they sold only two thousand of their first edition's twenty-five hundred copies, HKN surfaced through the process as a capable, energetic marketer, capitalizing on her well-honed organizational and secretarial skills.

A photograph taken of HKN at a promotional gathering at Macy's and later reproduced in their *Good Life Album* epitomized her new role. She stood smiling in the center of the image, skirted by copies of *The Maple Sugar Book* and holding a plate of sugar-on-snow. Sugar Association officials in business suits flanked her as they tasted the dessert. The setting was Herald Square rather than rural Vermont as

HKN flaunted the Nearings' experiment in simple living amidst the opulent, market-driven space that she and Nearing claimed to eschew. Macy's had been a favorite former haunt of Helen's and proved to be a lucrative account for their business, further illustrating an inherent tension between their "no-money" commitment and their burgeoning capitalist venture. Nearing hovered privately in the background of the photograph, barely visible, stirring hot syrup with his head turned away from the crowd, demonstrating his preferred distance from the promotional side of their work.[66] An unpublished promotional photograph taken of HKN at Forest Farm that same year, posing with a strained smile behind sugar jars, perhaps denoted an initial discomfort that would fade quickly as she flourished in her marketing career.

In *The Maple Sugar Book*, HKN emphasized the dailiness of their homesteading life and offered a handbook for potential followers who likewise might turn to a rural landscape as an alternative to Cold-War, consumer-driven American culture. HKN diffused the stridency of Nearing's less imitable radicalism through her more accessible writing style. Scott Skinner, the couple's grandnephew, later distinguished HKN's writing from Nearing's: "Scott had the writing style more of an economics professor, and Helen brought a wit and playfulness and sparkle to their joint projects that made them more interesting, and therefore reached a broader audience."[67]

HKN became conscious of her growing audience, having a natural inclination toward cultivating a public persona. Scott had written to Helen in December 1928 regarding this trait, "The crowd intoxicates you. You go to their level and meet their demands."[68] Her neighbor in Maine, Jeanne Gaudette, likewise described HKN: "I think the violin was simply replaced by another medium . . . [Helen] was a public person."[69] Publication and promotion of *The Maple Sugar Book* afforded HKN a new audience, with Macy's exhibits having a theatrical quality and Forest Farm becoming her natural stage for their good life model.

CHAPTER 4

Living the Good Life

Harborside, Maine, and Beyond ❧

TWO YEARS AFTER the publication of *The Maple Sugar Book,* Helen and Scott Nearing moved from Jamaica, Vermont, to Harborside, Maine. They felt compelled to relocate in light of increased development in the Pikes Falls Valley, particularly in the form of a proposed Stratton Mountain ski area.[1] They also were burdened by excessive numbers of visitors, and the maple sugar business had become too labor- and time-intensive.[2]

In 1951, the Nearings began parceling off their Vermont property in preparation for their move, deeding seven hundred acres to the town of Winhall to be preserved as a municipal forest. On January 4, 1952, the Nearings sold fifty-five acres of the Hoard land, including their large stone house, to George and Jacqueline Breen for $15,000, a price well below market value at the time. Nearing described this sale as another instance in which he was able to avoid the "menace of riches" by not exploiting a land sale. However, previous Vermont property transactions had garnered significant profits. Two years later, the Nearings sold their remaining 215 acres of land in Vermont to Pearl Buck. HKN returned in 1953 to guide the Breens through the sugaring season, after which she ceased sugaring altogether.[3]

An accomplished dowser who had found water on their two Vermont farm sites, HKN determined the region in which she and Scott would reside in Maine by running a divining rod over a map in 1951. HKN repeatedly honed in on Penobscot Bay and claimed that she "dowsed herself into exile" as they moved to Maine with "bag, baggage and a ton and a half of organic compost."[4] HKN lamented leaving the property they had cultivated in Vermont, as well as her family. Her mother, Maria Knothe, recently had moved to a nursing home in Manchester, Vermont, in order to be close to Helen and her siblings.

On October 19, 1951, HKN purchased a 140-acre uncultivated farm and drafty, wood clapboard farmhouse from Mary Stackhouse for $7,500. The property was located just across Goose Falls Bridge in the town of Harborside, Maine, on Cape Rosier.[5] It had rich, loamy soil, a vibrant freshwater spring, and a westward-facing section with dramatic views over Orr's Cove and Penobscot Bay.

The Maple Sugar Book had hoisted HKN into a new role as full partner and co-author in their homesteading venture. The pastoral spirit of her book spoke to post–World War II readers in search of physical and spiritual freedom. The 1950s in the United States was a time of burgeoning material wealth and sprawling suburbs. Nostalgic elements of pastoralism, as espoused in *The Maple Sugar Book,* expressed a need to conserve rather than consume, to retreat to natural landscapes as utopic alternatives to the dystopic city.[6]

Shortly after purchasing their property on Cape Rosier, the Nearings established the Social Science Institute, an independent, nonprofit press authorized to print and publish texts, to maintain a library, and to carry on other educational activities. They designated Nearing as president of the Social Science Institute, HKN the secretary, and appointed a group of like-minded friends and supporters as their advisory board. Through their Social Science Institute, the Nearings copyrighted their books and also offered seminars and lectures on topics ranging from world events to composting. Their students customarily found lodging at nearby Hiram Blake Camp or tented on adjacent properties. HKN offered tours of the Forest Farm garden, fed the SSI participants meals in their trademark wooden bowls, and put folks to work.[7]

In 1952, Maria Knothe's health began to fail. Though Helen's relationship with her stern father had been strained, she maintained a closeness with her mother that had become increasingly pronounced. In an exchange of affectionate letters, Helen wrote to her mother, "I feel so incredibly close to you. These last years and even months and weeks have brought us very close. I love you and feel a very real part of you." HKN wrote regarding the dying process, "Keep close contact with the inner light and source of life. Then you will live beautifully right through the 'death' phases, whether they come soon or late."[8] Helen related to Maria Knothe what Scott had told her regarding his mother's death: "There's some special link with a mother that gives added poignancy when they go." Finally, Helen described Maria Knothe's profound contribution in this life, "You have a certain light and sweetness and gentleness that radiates from you and that is yours alone. Treasure it and shine it. All light is needed in this world."[9] In Maria Knothe's last birthday letter to her daughter, written in February 1953, she wrote in a scrawling script, "I think of you every day and send you good wishes," signing the letter, "Love to you my dear, you mean so much to me, M."[10] Maria Knothe died later that year at age eighty-two. Any previous tensions with HKN's family faded in these final exchanges.

At this time, the Nearings also were engaged heavily in writing a report on their Vermont experiment. They completed two companion texts in 1954: a treatise by Scott Nearing entitled *Man's Search for the Good Life,* which provided the theoretical basis for the second, co-authored book, *Living the Good Life: Being a Plain Practical Account of a Twenty Year Project in a Self-Subsistent Homestead in Vermont, Together with Remarks on How to Live Sanely and Simply in a Troubled*

World.[11] *Man's Search for the Good Life* epitomized Nearing's erudite, anti-modern, polemical style as he outlined the ongoing search for the good life in a deeply troubled Western world. He put forth "A Plea for Social Sanity" in the midst of widespread pessimism borne out of:

> two increasingly total wars in one generation, the boredom of machine-tending, the senseless multiplication of meaningless gadgets, economic insecurity, the production and stockpiling of atomic weapons, the feverish search for more efficient ways of destroying property and crushing out life, the enlarged power of the military and the ceaseless, war-engendered propoganda of fear.

Nearing argued that the Cold War was a struggle for "job ownership, public opinion and police power."[12] His critique contradicted prevailing notions of American prosperity as rooted in burgeoning modern conveniences, baby booming, two-car families, and Cold War patriotism, all of which he believed obstructed one's search for an authentic good life.

Change must happen on individual and collective levels, according to Nearing. He cited Confucius, Tolstoy, Socrates, Jesus, Gandhi, and Lenin as exemplars who opposed existing social systems. Nearing called for social change through a rejection of the habits and traditions that inhibit conscience and reason. "Those who propose to live satisfying, productive lives from now to 1999 A.D. must expect to do their share of pioneering," wrote Nearing.[13]

Living the Good Life differed from *Man's Search for the Good Life* in that it was co-authored with Helen and was autobiographical. The Nearings described daily life on their Vermont homestead, combining theory with practical, personal examples. The Nearings characterized their move to Vermont as a conscious withdrawal from American culture, which they saw as "competitive, acquisitive, predatory."[14] They claimed to have embarked upon their "twentieth century pioneering venture" with three primary objectives in mind: to make themselves as independent as possible from commodity and labor markets; to improve and maintain good health through consuming homegrown organic food and closer contact with the earth; and to liberate and dissociate themselves from exploitation such as "the plunder of the planet, the slavery of man and beast, the slaughter of man in war and of animals for food."[15] Referring to their homestead as a "laboratory," they included detailed accounts of their organic gardening strategies, along with their building projects, their maple sugar business, and their strict vegetarian diet. The Nearings also touted their daily regimen, in which they divided their time into four-hour increments. The Nearings offered their text as a guide for the "many individuals and families, tied to city jobs and dwellings, who yearn to make their dreams of the good life a reality." They provided a handbook for homesteaders rather than a lofty, theoretical analysis.[16]

In *Living the Good Life*, HKN also included visual images. Documentary photography or "social seeing" had become an increasingly influential genre popularized in the United States by Lewis Hine in the 1930s and by James Agee and Walker Evans with *Let Us Now Praise Famous Men* in 1941. Recognizing the potential for publicity in photographs and the social currency of countercultural images, HKN masterfully made photographs a part of their good life message. She advertised *Living the Good Life* as being "with photographs" and made images of their homesteading work indelible symbols of their alternative life. Scott Nearing preferred words to images and would remain staunchly "more at ease with facts than photographs."[17]

Living the Good Life became the story that Helen in particular would internalize and retell in memoirs, speeches, and articles and would reinforce with photograph albums and scrapbooks. In this narrative, HKN highlighted the Nearings' close, nonviolent relationship with the natural landscape; their belief in organic food and architecture; and their commitment to seeking health through simple living. Her version of their story idealized their homesteading experiment, omitting truths about profitable land sales and obscuring facts regarding the extent to which others helped them with their project. Nevertheless, their good life narrative proved intriguing to a sizeable readership, and they sold ten thousand copies in the first sixteen years of its publication, attracting thousands of visitors to their remote farm in Maine.[18]

The Nearings' romantic representation of living on the land was characteristic of, and derivative from, a larger homesteading corpus, as homesteaders are known for "producing as many texts as vegetables."[19] Henry David Thoreau's *Walden,* which celebrated its centenary year of publication in 1954, was the "original sacred text on homesteading" but soon was followed by a plethora of others such as John Burroughs' *Signs and Seasons* (1886), Ralph Borsodi's *The Flight from the City* (1933), Gove Hambidge's *Enchanted Acre* (1935), and Louis Bromfield's *Pleasant Valley* (1945).[20] Like homesteaders before them, the Nearings echoed Thoreau's call for a simple life attached to the land, continuing in a well-established, American pastoral literary tradition and becoming spokespeople for simple living in the markedly eclipsed, mid-century American left.

Living the Good Life offered a "core vision of harmony" and a balm for a troubled world.[21] Songwriter and social activist Pete Seeger dedicated a song to the Nearings entitled "Maple Syrup Time":

> *I'll send this song to Scott and Helen*
> *Up in Maine where they are dwellin',*
> *Hoping they don't mind*
> *a little advice in rhyme.*
> *As in life or revolution,*
> *rarely is there a quick solution,*
> *Anything worthwhile takes a little time.*
> *We boil and boil and boil*
> *and boil it all day long.*
> *When what is left is syrupy,*

Don't leave it on the flame too long.
But seize the minute, build a new world,
Sing an old song.
Keep up the fire! Maple syrup time.[22]

MAPLE SYRUP TIME by Pete Seeger
© Copyright 1977 by SANGA MUSIC, INC.

His lyrics "sing an old song" and "keep up the fire!" summoned a call to any listeners to "seize the minute" and pursue alternative courses as had the Nearings.

The Nearings remained public figures, continuing to lecture and write after leaving Vermont. Between 1952 and 1955, they embarked upon three extensive cross-country American tours. They drove their station wagon fifty thousand miles during sixteen of the thirty-one months from October 1952 to May 1955. They funded their travels through lecture fees and the generosity of hosts who were primarily readers of *Monthly Review,* a socialist periodical for which Nearing wrote a "World Events" column from 1949 to 1972.

The Nearings traveled to forty-seven states, as well as portions of Canada and Mexico, conducting approximately six hundred meetings with more than thirty thousand people. No longer tied down by their maple sugar business, HKN accompanied Nearing on these lecture tours and assumed a position of equality and notoriety as both co-author of their *Maple Sugar Book* and long-term partner in their homesteading work. The purpose of these trips was to promote *Monthly Review,* to spread their anti–Cold War message, to publicize their good life alternative, and to educate themselves on American political perspectives under the shadow of the Korean War. The Nearings co-authored a book entitled *USA Today,* generated from these travels, in which they spelled out their position against the Cold War and the concomitant Red Scare:

The tide is turning and will continue to turn, as more and more American people realize how thoroughly they have been sold down the river by vicious propaganda which smothered their normal friendliness and human sympathy under a blanket of synthetically generated fear and hatred.[23]

Embedded in the Nearings' foreboding rhetoric was their resounding critique of the "synthetic" or inauthentic.

From the mid-1950s through the 1960s, the Nearings traveled internationally during the winter months, both in defiance of the ardent nationalism engendered by the Cold War and also out of concern that their passports—due to expire in 1958—might not be renewed. The Nearings embarked upon particularly ambitious journeys during the winters of 1956–1957 and 1957–1958. They crisscrossed Canada, spent three weeks in Japan, three months in Southeast Asia, and returned home via Damascus, Cairo, and Rome. In 1957, they attended an International Vegetarian Union Congress in India. Nearing had served on the Executive Committee of the IVU for many years. Again, they were hosted primarily by *Monthly Review* readers and other socialists from whom they gathered information for their co-authored book *Socialists Around the World* (1958).

In the winter of 1957–1958, the Nearings traveled to the Soviet Union and the People's Republic of China in order to visit the two countries "that were making the most extensive efforts to establish planned and purposed social order." The Nearings hoped to fortify their understanding of these countries. It had been twenty-one years since they had visited Russia and thirty years since Nearing had traveled to China. They also wanted to be in Russia for the fortieth anniversary celebration of the October Revolution. In their co-authored report on their trip, *The Brave New World,* they claimed that they had gone to observe the Russian and Chinese people as they worked and played, and, contrary to the thinking of most leftists in the West at the time, the Nearings believed that the New World was "changing for the better":

We bring good news to the people of the West. There is a New World growing up in parts of Europe and Asia. We have seen it, been in it and of it, watched it develop. Needless to say, the New World is different in many respects from the Old World. Its objectives are new. Its pattern is new. Its ways of life are new.

They argued that Russia and the People's Republic of China aspired to build peaceful socialist societies; ongoing hot and cold wars would stand in the way of such progress. The Nearings called for recognition of Russian and Chinese people as fellow world citizens.[24]

However, the Nearings' claim that the New World they encountered was changing for the better denied the realities of the human rights violations that the respective socialist governments had incurred. Their argument maintained Nearing's unflagging defense of the Soviet Union, a puzzling aspect of his dogmatism that otherwise was rooted in nonviolence and social justice. The Soviet Union even had outlawed vegetarianism in 1917. Yet Nearing was rigidly idealistic, denying realities that might contradict concepts that he embraced. HKN similarly was willing to suspend disbelief in support of their stories.

In the early 1950s, Nearing severed all ties with his son John because of their conflicting views regarding the Soviet Union. John Nearing had worked in the Soviet Union during the 1930s and married a Russian woman. He became disillusioned with Soviet policies during the Moscow Show Trials of 1936 to 1938 and, in 1942, left the Soviet Union to work for *Time Magazine,* founded by Henry Luce, whom Nearing felt epitomized capitalist depravity. Nearing eventually ended contact with his son over their political differences. As the Moscow trials, the Hitler-Stalin pact, World War II, the ensuing Cold War, and the Soviet repressions of Hungarians and Czechs dissuaded many Soviet sympathizers of his time, Nearing consistently ignored or defended Soviet injustices and human rights atrocities, admiring Tolstoy and Stalin equally.[25]

The Nearings likewise lauded the Chinese Communist Party and Mao Zedong, who had come to power in 1949 and had reached their zenith of control over the Chinese mainland at the time the Nearings arrived in 1957. That same year, Mao Zedong unleashed his Anti-Rightist campaign, suppressing all counterrevolutionary enemies and ushering in a period of social repression and economic stagnation that would persist until 1989.[26] Yet Nearing remained an ardent apologist for both the Soviet and Chinese governments, and HKN likewise wrote on their behalf.

Both Nearings were also long-term, ardent pacifists. They were staunch critics of America's virulent anti-communist containment policy born out of the 1947 Marshall Plan and the subsequent document NSC-68 (1949), which mandated "massive military buildup at home and . . . situations of strength abroad." The Nearings adamantly opposed this antagonistic stance toward the Soviet Union and the ensuing establishment of NATO, the onset of the Korean War, and the increased U.S. military involvement in Southeast Asia.[27] In *Socialists Around the World,* they wrote that "the desire for peace is uppermost in the minds of hundreds of millions" and quoted Albert Einstein's proclamation, "We must disarm or die."[28]

Throughout the sixties, the Nearings continued to travel, informing their lectures and pamphlets while also satisfying HKN's insatiable wanderlust. They returned to China, traveled to Cuba, and conducted a lecture tour through Europe. Nearing traveled on to South America, meeting with communist party officials in each locale. During the winter months, HKN sought reprieves in Florida or met Nearing at the Rio Caliente spa in Guadalajara, Mexico. In April 1963, they traveled together through Guatemala, though their wanderings were curtailed later that year as their passports were revoked temporarily because of their suspected communist ties.[29]

In January 1965, with renewed passports in hand and sufficient revenue from books and lectures, the Nearings visited Bombay and then returned to India two years later for an International Vegetarian Union Congress and a meeting with the president of India in Delhi. They toured Ceylon, Hungary, and Israel as well, participating in international vegetarian gatherings, communist party meetings, and peace talks. With the American involvement in Vietnam escalating, Nearing publicly condemned the United States' activities there, describing the United States "as an animal slaughtering, man, woman and children slaying, killing nation."[30]

Despite their condemnation of American foreign policy, the Nearings chose not to become expatriates. Instead, they looked to their farm in Maine as a retreat from global strife and invited listeners and readers to visit them there in order to see and experience a working, positive model of a potential "fresh start."[31] In a January 18, 1961, letter to Helen, Scott at age seventy-seven outlined his life plan:

> 1. I am going back to my post in USA. Continuing with Forest Farm, SSI, and the rest of the job of subsistence and education. This for at least several years.
>
> 2. As the USA in particular and the human race in general moves closer and closer to a general war of annihilation, I will work against war and the war system, for coexistence and cooperation.
>
> 3. In tactical matters I will work against fear, hatred, violence; for understanding tolerance, patience and persuasion.[32]

The Nearings' Maine homestead remained not only a place of subsistence and retreat, but also, as Rebecca Kneale Gould described it, an "outpost from which to continually project their manifestos of cultural criticism into the public sphere."[33]

To allow adequate time for writing and travel, the Nearings at first "camped out" in Harborside, not beginning "serious living there until 1954–1955."[34] They avoided elaborate renovation of their farmhouse, adding only a fireplace, a balcony, and a woodshed. Their early efforts in Maine were directed mostly toward growing and harvesting blueberries as their cash crop, cultivating their one-quarter-acre garden and, over the course of fourteen years, building a five-foot-high stone wall around the perimeter of their garden. They also dug a half-acre spring-fed pond for swimming and skating, hauling fourteen thousand wheelbarrow loads of clay.[35]

HKN collected numerous telling photographs of their life in Maine: cellar shelves teeming with jarred goods, an outdoor table for shared meals, hand plows, a lush herb garden, meticulous wood piles, and a sun-filled greenhouse. The photographs also depicted HKN's few decorative touches: a half wagon wheel set over their front door, mock outside shutters painted beside windows, kitchen cabinets with Dutch details, and favorite postcards taped or thumbtacked to walls. The Nearings' living space in Maine was readily conducive to writing and mailing, as their bookshelves and file boxes were set in close proximity to their dining and work table. HKN's kitchen was efficient and organized, with dried herbs draped from the ceiling and wooden bowls and utensils on open racks beneath which hung wooden stamps for mailings.[36] HKN continued to limit food preparation by designing simple dishes based on raw fruits and vegetables. She devoted scant energy to house cleaning, instead directing her physical labor to the garden and building projects. She enlisted help from visitors for household chores.

Their lecture tours and the 1954 publication of *Man's Search for the Good Life* and *Living the Good Life* attracted numerous guests to Harborside despite the Nearings' remote locale. The Nearings remained exceedingly hospitable, opening their home on a daily basis. However, in order to protect their time, HKN hung a sign at the end of their driveway requesting that visitors help them live their good life by coming only between the hours of three and five.[37]

HKN particularly was engaged with the guests, despite her penchant for solitude and her impatient nature. Visitors were attracted to HKN's energy as she tempered Nearing's austerity and rigidity by "skipping from rock to rock," yodeling, skating on their pond, and playing music. HKN would decorate a straw figure at harvest time in traditional Dutch clothes and create fanciful snowmen. Nearing tended to be more reticent, whereas HKN would converse freely with guests. She typically did not engage in casual, light conversations, however. Talk was work-related or might refer to pressing current is-

sues. A visitor described upon first meeting HKN having a "little argument" about Albania and communism and whether or not China
was truly a communist country. HKN stayed abreast of current events
and "cultivated the life of [her] mind and spirit" with extensive reading on topics ranging from politics to world vegetarianism to UFOs.[38]

HKN also demonstrated a profound spiritual connection to their
natural landscape, offering a syncretic homesteading model that
blended the mystical with the practical. She remained an avid sun
worshipper, still believing in the therapeutic power of rays and in
metaphorical notions of light as representing a creator or guiding
force. Her collection of photographs included pictures of gardening
and building interspersed with spiritually redolent images of seagulls
or heart-shaped rocks that she believed were imbued with mystical
powers. Both she and Nearing wrote of a continued reliance upon
séances and meditation to direct their thoughts and actions, and
HKN might offer a wishing stone to a child or casually read a friend's
palm.[39]

Through her flouting of mainstream Cold War nationalism, her
practical work close to the land in Maine, her ongoing hospitality, and
her rich spirituality, Helen Knothe Nearing increasingly found her
purpose in their homesteading life and surfaced as a particularly
effective, intriguing purveyor of the Nearings' good life ethic.

CHAPTER 5

Back to the Land

The Environmental Movement and
a Final Forest Farm ✻

IN THE MIDST of a post–World War II boom economy, the natural environment became a sought-after amenity within an evolving ethic of consumption. However, increased pollution, a byproduct of the boom economy's industrial growth, soiled rivers, lakes, and air, diminishing the quality of Americans' outdoor experiences. In 1948, the first modern anti-pollution law, the Water Pollution Control Act, was passed, but was then followed by a decade of inadequate, non-confrontational solutions to environmental decay.[1]

The early 1960s marked a notable shift in attitude from earlier conservationist sensibilities to ardent environmentalism as ecological crises became increasingly acute.[2] The Clean Air Act of 1960 represented a significant legislative effort, and in 1962, Rachel Carson's dire warning against the indiscriminate use of DDT (dichloro-diphenyl-tricloro-ethane) and other pesticides in her book *Silent Spring* raised cultural consciousness, spawning the late-twentieth-century American environmental movement.[3] Helen and Scott Nearing had issued a similar warning against the use of chemical pesticides, in particular DDT, in *Living the Good Life,* published almost a decade before *Silent*

Spring.[4] Environmental sensibilities finally were catching up to the Nearings as public values shifted from resource and wildlife conservation to concern for human impact on the earth and the concomitant effects on the health of people and the planet. Pastoralism and its nostalgia for open spaces, clean air, and fresh water began to assume political significance and have progressive implications.[5]

The environmental movement was characterized on a local level by increased recycling, organic gardening, and avoidance of domestic toxic chemicals.[6] Membership in environmental groups increased throughout the sixties and seventies, and environmental legislation such as the Water Quality Act (1965) and the National Environmental Policy Act (1969) were passed in response to disasters such as the polluted Cuyahoga River's bursting into flames and the Santa Barbara oil spill.[7] The first Earth Day celebration on April 22, 1970, marked a pinnacle in public support for environmentalism in the United States.

Back-to-the-landers' countercultural search for alternative lifestyles went hand in hand with this environmental fervor. A rural life of voluntary simplicity promised not only self-sufficiency and a reduction of one's economic needs through a spiritual commitment to "enoughness," but also a way of life that would be environmentally sensitive and sustainable.[8] A number of periodicals such as *Mother Earth News* (1970), *The Whole Earth Catalogue* (1974), *Rain* (1974), and *Organic Gardening* (1948) provided philosophical as well as practical support for the movement, as did new courses in organic farming offered at agricultural colleges. J. I. Rodale, founder of *Organic Gardening* and the individual credited with having introduced the American public to organic farming techniques, was even interviewed for the popular Dick Cavett television show in 1971. Tragically, Rodale died during the Cavett interview, and his son, Robert Rodale, thereafter became the primary spokesperson for *Organic Gardening.*[9]

Writers such as Barry Commoner and Wendell Berry likewise offered critiques of pesticide-laden commercial agriculture, calling for organic soil management and recognition of the small organic farmer as a model "nurturer." Berry, a "new homesteader" himself, character-ized such farmers as environmental pioneers: "people of principle, both stubborn and adventurous, independent enough to trust their own experience and strong enough to hold in considerable isolation to truths not officially or popularly favored."[10] These new home-steaders typically were less politically involved than mainstream en-vironmental activists. They were instead "totally committed to their pastoral vision."[11]

Helen and Scott Nearing surfaced as the elder statespeople of this back-to-the-land movement. They offered a home-based environ-mentalism and exemplary model of successful homesteading for these pastoral idealists.[12] New homesteaders were captivated by the Nearings' good life story, and, in response to popular demand, *Liv-ing the Good Life* and *The Maple Sugar Book* were republished by Schocken Books in 1970.[13]

A number of other, older, environmental texts likewise experi-enced a rebirth at this time. Aldo Leopold's 1949 *A Sand County Al-manac* was reprinted in 1966 and 1970, and Richard Gregg's 1934 pio-neering pamphlet on organic farming, *Companion Plants and How to Use Them*, was republished in 1966 and 1973.[14] However, as one book reviewer noted, the Nearings' books proved unique in that they be-came the "bibles of the present back-to-the-land counterculture."[15] *Living the Good Life* sold over fifty thousand copies in its first year of republication and attracted more than two thousand annual visitors to Forest Farm over the course of the next decade.

Glowing reviews appeared in periodicals such as *Freedom News*, the *Catholic Worker*, the *Nation*, and *Natural Life Styles*.[16] *Living the Good Life* and *The Maple Sugar Book* also were listed as suggested "Christmas Parcels" in a December 1971 *New York Times Book Review*. Less favorable reviews appeared in *Time*, *Newsweek*, and *Harper's Magazine*, but those articles for the most part criticized the counter-cultural followers of the Nearings rather than the Nearings them-selves.

Scott Nearing—with his seasoned radicalism characterized by a strenuous, disciplined lifestyle and steeped in socialist theory—likewise disparaged the "sexual, sensory hedonism" of their new followers.[17] He chastised hippies for privileging freedom above all else, and for their poor hygiene, abuse of drugs and alcohol, and profligate use of time.[18] HKN, in a 1979 interview, described the "seekers" who came to their Maine farm:

> Almost universally they favor "freedom": that is the pursuit of their personal goals and fancies. They are not joiners and generally not members of any group more specific than is implied by the adoption of a specific diet or the practice of some yoga exercises. They are wanderers and seekers . . . Perhaps they can best be described as unsettled. Never before in our lives have we met so many unattached, uncommitted, insecure, uncertain human beings.

HKN's harsh critique proved ironic, given that both she and Nearing likewise sought personal freedom, did not belong to any political groups, and attended off-beat vegetarian conferences. Further criticizing the demographics of their guests, HKN claimed as of 1979 not to have had a single black visitor, nor a daughter or son of a coal miner, factory or mill worker. Their visitors were for the most part middle-class, privileged youth—as the Nearings had been—whom HKN described as "born into affluence of a sort, raised in comfort" with "little or no work experience."[19]

Historian David E. Shi concurred, in *In Search of the Simple Life* (1986), that many young, rebellious back-to-the-landers were simply unable to emulate the Nearings' successful practice because they were "woefully unprepared for the requirements of living on the land." Rural hippie settlements tended to last only a few months or years at the most.[20] Nevertheless, the Nearings became the ideal to which the back-to-the-landers aspired. Keeping up with the Nearings became almost "another form of the rat-race."[21]

The Nearings, despite their latent criticisms of these "seekers," offered numerous visitors the opportunity to see and participate in their homesteading life. They modeled their good life through ongoing work and discipline, lighting a wood stove each morning, growing their own food, composting, recycling, and teaching. The Nearings had no television, radio, or telephone, but kept abreast of current events through extensive reading of books and periodicals. HKN was "ahead of her time" with respect to recycling. She rarely sent an envelope or note to a friend that was not reused from a previous mailing.[22] "They were living their convictions," contended their Maine neighbor Gail Disney, and the Nearings required that their visitors live them as well.[23]

The Nearings continued to herald hospitality as central to their homesteading life. According to HKN, all visitors were "possible recruits for a general effort now under way to stablize and improve man's earthly living space . . . [and] to raise popular interest and determination to the action point."[24] Both Nearings "were perpetual teachers," claimed a former visitor, as guests still were invited to attend Monday night discussions, to sit in on Social Science Institute lectures, or to join Sunday night music gatherings.[25]

Nancy Richardson Berkowitz began working closely with Nearing as a gardening apprentice in the early 1970s. She also helped HKN welcome guests to Forest Farm and described Helen's extraordinary willingness to "open her house to total strangers every single day," even though the number of visitors could range from one or two to a dozen or more.[26] If the Nearings were engaged in a task when people arrived, they continued working and invited the guests to join them. If they were eating a meal, all visitors were given wooden bowls and spoons and served as well.[27] The guests were offered standard Forest Farm fare: leafy vegetables and sprouted grains; uncooked rolled oats, oil, and raisins; boiled wheat or millet; peanut butter and honey ("Scott's emulsion"); "carrot croakers" (journey cakes with grated carrot base); or lots of apples. They "had worthwhile and happy times," according to HKN, and some of the visitors became

"very good friends, corresponding and coming back at every opportunity." Nancy Berkowitz remembered, "It was fun. We had every-
body," and she described guests arriving from as far away as Alaska
and England, appearing on foot or on bicycles.[28]

As the primary coordinator of these guests, HKN not only organized gatherings and prepared meals, but also capitalized upon this
free labor force for the running of their homestead. She broke visitors
up into work teams according to "inclination and experience." She
later described them: "With these crews, we could dig foundation
trenches, pour concrete, cut brush, and above all build roads. People
who had skills exercised them. Others of less experience learned."[29]
One guest explained that the Nearings' "systematic approach was evident in everything I saw." Visitors were expected to adapt to the
Nearings' routine, participating fully in their arduous work.[30]

"Some few [guests] had real skills," HKN remarked regarding those
visitors who had worked previously on farms or building projects. A
handful of the "seekers" went on to become successful back-to-the-
landers with their own flourishing properties. HKN would warn aspiring homesteaders that they needed both start-up funds and help
with the labor because homesteading work was too much for just an
individual or a couple alone.[31] In the late sixties, as Nearing approached his ninetieth birthday, the Nearings initiated plans to build
a final, smaller homestead. In turn, they began parceling off their 140
acres, selling portions "to the right young people."

The Nearings saw Eliot and Sue Coleman as a "promising young
couple" in 1968 and sold them sixty acres of their Forest Farm property.[32] In 1972, the Nearings sold twenty-two more acres of their property to Jean and Keith Heavrin, who "went in for animal husbandry
and did construction work," as well as a small piece to Greg Summers,
who "raised a good garden in the woods and took print-shop jobs to
bring in the necessary cash."[33]

These back-to-the-landers were not ardent socialists, nor were
they necessarily well versed in the Nearings' economic or political

theories, beyond basic aphorisms such as "pay as you go."[34] Instead, they were inspired by the Nearings' autobiographical texts and their ongoing example. Helen, in particular, proved more accessible than Scott to this younger, countercultural generation. Her eclectic mysticism paralleled the New Age spirituality to which many of her youthful followers adhered. HKN's spiritual connection to the natural world—as expressed through meditation, yoga, astrology, Ouija board sessions, and contemplation in the peaceful sanctity of their wooden yurts—resonated with late-twentieth-century notions of cosmic connections to the earth as a living organism.[35] HKN's theosophical background also held appeal when Krishnamurti and theosophy experienced a resurgence in popularity during the sixties and seventies.[36]

The reissuing of *Living the Good Life* and *The Maple Sugar Book* ushered in an extremely prolific period for the Nearings. They wrote a regular column for *Mother Earth News* throughout the 1970s, while also adding eight more volumes to their corpus.[37] In 1972, Nearing completed his fiftieth book, *The Making of a Radical: A Political Autobiography.* Though dedicated to Helen, "who did half the work," *The Making of a Radical* referred only sparingly to HKN and their homesteading practices. Instead it traced the evolution of Nearing's political thoughts up to his eighty-ninth year.

Two years later, HKN published *The Good Life Album,* her first book as a sole author. A vastly different memoir from Nearing's, *The Good Life Picture Album* was a photographic compendium of the Nearings' lives organized in backwards chronological order. After a brief introduction summarizing their Vermont and Maine experiments, HKN visually expanded their story beyond homesteading, tracing back to their familial roots and spanning the ninety-one years of Nearing's life and seventy of her own. She ended the text by coming full circle with an elderly shot of herself sitting beside a grinning, wrinkled Nearing, an unusual reprieve for the industrious couple and a cheerful denouement.[38] For HKN, who assiduously compiled pho-

tograph albums and scrapbooks and introduced pictures as part of the
Nearings' good life story, this text marked a natural first solo book attempt, particularly given that she never perceived herself to be a "great
writer."[39] Rather, she was a skilled marketer and publicist, and this text
became a popular addendum to the Nearing corpus as a visual autobiography and, ironically, a glossy rendition of their simple life.

HKN's promotion of the visual aspects of their story was reinforced by a 1977 documentary of their good life produced by Bullfrog Films. Images of their garden, their building projects, Helen yodeling at dusk in their tranquil cove, and visitors listening to Nearing
lecture were interspersed with the Nearings' narration of the history
and present status of their homestead. With a eulogistic tone, the documentary commended their efforts and reinforced the Nearings'
good life narrative with moving pictures.[40]

In 1977, the Nearings co-authored *Building and Using Our Sun-
Heated Greenhouse,* in which they highlighted their greenhouse and
cold frames as affordable, practical means for growing year-round
vegetables. Two years later, Schocken Books published the Nearings'
final co-authored text, *Continuing the Good Life* (1979). A sequel to
Living the Good Life, the text offered a detailed account of their Maine
homesteading experience.[41] The Nearings described the challenges of
Maine's short growing season, winter gardening in their greenhouse,
as well as winter food storage. They contended that blueberries were
their Maine cash crop, although they never successfully generated an
income from blueberries. The Nearings touted building with stone
and described the four-hundred-and-twenty foot stone wall that they
had built around their garden.[42] They credited their sustained health
to their diet and physical activity and emphasized in *Continuing the
Good Life* the arduous work they continued to do despite their advanced ages.

Both Nearings were paragons of well-being and independence in
later life. One *New York Times* book reviewer contended that "the
Nearings' skill lies in describing this half-century of deliberate labor

in such a way as to make it intensely appealing to a whole segment of American society," old and young alike who were looking for alternative ways to live.[43] The review, entitled "Scything the Meadow at 95," described the Nearings as together having "joined, almost founded the counterculture" fifty years prior and as living it into old age.

In 1980, HKN wrote a cookbook, *Simple Food for the Good Life*, which outlined their good life diet. She offered a collection of basic recipes and quotations about food and cooking, as well as a critique of typical unhealthy American fare.[44] HKN claimed in *Simple Food* to have been asked to write the book in response to their growing readership but noted the irony of compiling a cookbook when her primary culinary goal was not to spend time cooking. She imparted the benefits of vegetarianism, raw food, and electric blenders along with remarkably bland recipes. Joe Allen, in a review for *Ad Vantages* magazine, cited HKN's odd recipes for "Horse Chow," "Haymaker's Switchel," and "Squash Cutlet," urging readers not to dismiss the strange dishes in light of the Nearings' longevity and excellent health.[45]

HKN also published *Wise Words on the Good Life: An Anthology of Quotations* (1980) in which she compiled hundreds of the quotations she had collected. With a penchant for old books and bookshops, HKN had culled sayings from a variety of texts and in *Wise Words* organized those quotations into chapters on country life, health, labor, gardening, building, simple living, hospitality, solitude, and old age.

HKN gained further recognition as purveyor of their good life ethic through the physical space of their final homestead. In 1971, they began construction on another chalet-style, stone house that would be a "sister house" to the one in Vermont. HKN described the project as primarily her own: "I was the instigator, the planner, the chooser of the site, the designer, the interior decorator and through a kind of legacy from an old Los Angeles friend, the one who paid the bills."[46] HKN selected the portion of their property overlooking Orr's Cove, which she called "Spirit Cove," for the building site. Scott protested that the northwest winds would prove too cold and the soil too soggy, but Helen would not relent because the location offered a stupendous, mystical western view over Penobscot Bay.[47]

The Nearings remained stubborn and opinionated as they grew older, but they also developed a particularly deep, mutual regard for each other. During an interview late in life, Nearing was asked, apart from his heroes Tolstoy and Gandhi, what contemporary person most influenced him, to which he responded, "Helen."[48] Their Maine neighbor, Jeanne Gaudette, described them as complementary "astrological opposites":

> They were soul mates . . . But they were different, they were really different. He was deliberate, slow and had few words and . . . she would bounce off the walls . . . Like when she would go, "Oh, I have an article for you," she'd leap up, and she might very well smash her hip up against the table . . . She was like a rogue neutron. Really, she was very eccentric that way.[3]

According to Gaudette, Nearing described the 1970s as "Helen's time to shine." Nearing began deferring to HKN, who would "interject a bit in talks, and would assist him much more" and who made most decisions related to the construction of their final Forest Farm.[49]

Following HKN's basic design, Brett Brubaker served as the principal architect and builder on the project.[50] In Vermont, the Nearings had built with prime granite, whereas in Maine, HKN opted to use colorful stones collected primarily on their nearby beach. In 1971, workers began clearing the building site, and the first stone building, the outhouse, was constructed on the property in 1973. The garage/workshop was built in 1973 and 1974, at the same time that the cellar for the main house was excavated. The Nearings dedicated the summers of 1975, 1976, and 1977 to their building project and were able to move into their new home in June of 1978.[51]

HKN's fourth book as a sole author, *Our Home Made of Stone: Building in Our Seventies and Nineties* (1983), was a photographic journal depicting the construction of this final homestead. She introduced the project by proclaiming the Nearings "joy in birthing new buildings," having built more than thirty stone structures in Vermont and Maine, ranging from outhouses to garden walls to sizable homes. She described their belief in Frank Lloyd Wright's organic architecture as growing from "sheer internal necessity"in contradistinction to arranged and artificial architecture, which has been defined as 'the manufactured house set in the midst of a manufactured environment.'" HKN claimed, "We dared to believe that ordinary, untrained people could make their own homes as birds build their own nests," in the process promoting their good life and their environmental idealism.[52]

A utopian reviewer wrote regarding the Nearings' house as depicted in *Our Home Made of Stone:* "Absent is the petty, the useless, and the mean. The walls and the furnishings speak of utility, freedom, and the warmth of generosity."[53] Absent also was Brett Brubaker's name. HKN included Brubaker in photographs but failed to give him due credit as the architect and builder. The book displayed a collection of photos that rendered Helen Nearing, rather than Brubaker, as primarily responsible for the construction of the house, and her introduction likewise glamorized and embellished her own work on the project, downplaying the contribution of others while reinforcing the Nearings' good life narrative.

Nevertheless, their final Forest Farm did offer a climactic distillation of their good life message, as well as insight into HKN through its design and artifacts. Stone walls were bolstered by hand-hewn wooden beams and rafters; interior walls were pine paneled; and Helen's three favorite words, "birdsong, trees, snowfall," were carved at her request into a knotty-pine board in the stairwell. Indoor plumbing, electricity, a telephone (added shortly before Nearing's death), and electric appliances denoted separation from and impact upon the landscape. Each room opened up onto the natural beauty of their wooded, seaside setting through oversized windows that were aesthetically appealing but offered scant insulation in the winter.

The basic form of the house—Alpine chalet-style, elegant stone walls, the main living room's grandiose picture window looking out on Penobscot Bay—had a "connector" quality, and its simple beauty attracted residents and visitors. However, this attractiveness was offset by a cold, interior flagstone floor, uncomfortable furnishings, and inadequate heat sources. The basic form connected, while the core separated. As Nancy Berkowitz explained: "The house is inviting in a certain way and in another way, it doesn't want you to stay too long."[54]

HKN's interior decorations were sparse and revealing. She did not add color with curtains or rugs but used only a bamboo shade for the large picture window and an industrial brown carpet upstairs. However, with a nod to her Dutch heritage, she painted the ceiling of Nearing's room a bright Van Gogh blue. She again thumbtacked favorite postcards to walls, hung Japanese and Chinese art, along with photographs of their families and images of children skating in Holland.

The living room was lined with overflowing bookshelves. HKN's Theosophical Society plaque, reading "There is no religion higher than truth," was placed amidst the spiritual texts. Celestial-looking globe lights hung from the ceiling to create an ethereal atmosphere. HKN felt a particular affinity for bells, which she placed inside and out, and she likewise displayed images of cats, Buddha statuettes, Japanese hats, and Dutch ice skates. By the outside dining table, near where Helen would kick off her Dutch wooden shoes or flip flops, the Nearings kept a basket of mismatched gloves that their workers could use. Nail and potato buckets in the pantry held bulk foods such as wheat or oats, a variety of teas, vegemite, and plastic bags to be reused. A dictionary, to which both Nearings referred frequently, was kept next to the kitchen table. HKN later would cover a small section of the kitchen wall in pictures and postcards of red poppies to commemorate a favorite poppy from her garden, though freesias were her favorite flowers. She also created a blue heron wall at the top of the stairs as a memorial to a blue heron that spent a great deal of time in her yard shortly after Nearing died. HKN believed that the bird was either a messenger sent by Nearing or perhaps a spiritual incarnation of Nearing himself.

The material artifacts of Forest Farm exhibited the Nearings' good life values. HKN chose to have a Swedish Clivus Multrum self-composting toilet or "earth closet" installed and would use the liquid and solid compost from their toilet as fertilizer on the property. They burned the wood that Nearing chopped for fuel in their stoves and continued to eschew china and cutlery, eating with chopsticks, wooden spoons, and wooden bowls. Dried herbs and vegetables hanging from the kitchen rafters reinforced their vegetarianism, and HKN decorated a wall of the pantry with pictures of slaughtered cows and multipronged fish hooks in order to shock any meat-eaters on the premises.[55] Possessions such as an electric space heater revealed pragmatic, day-to-day needs, as the house not only exhibited the Nearings' good life values but also functioned as a living space.

The exterior landscape included a stone garden wall, stone greenhouse, and stone garage that were integrated parts of the whole structure. In 1972, a Nearing friend and the founder of the Yurt Foundation, William S. Copperthwaite, organized a group for a two-week retreat to construct a yurt in the woods behind the Nearings' garden. He arranged the building of a second yurt fifteen years later.

In turn, the homestead offered HKN's material expression of their story, a final stage for their good life enactment and an artifactual autobiography of Helen Knothe Nearing. A visitor described her impressions of Forest Farm in a 1993 letter to Helen: "For me it was like the pages of the Nearing books jumping out at me in living color . . . Forest Farm certainly stands as a material symbol of the ideals and convictions that Mr. Nearing held, and you still hold, as a living example for the rest of us."[56]

In 1980, HKN sold their old clapboard house to Stanley Joseph for $75,000, ten times what she had paid Mary Stackhouse for the 140 acres in 1952.[57] HKN justified her profits from this sale: "[Stan] had the money, and we could use that money for this [new] house. We were not as pure in Maine as we were in Vermont."[58] Vermont land sales, however, also had not been "pure" or devoid of profit. Their back-to-the-land existence continued to require revenue.

Though HKN had taken charge of the sale of their old house and the development of their new homestead site, Nearing, at age ninety-seven, still remained a hearty contributor to their work. A remarkable example of the health benefits of their good life regime, he regularly chopped wood and raked compost until 1981. He and HKN had continued traveling and speaking nationally and internationally throughout the seventies and eighties. In 1971, they traveled to the Hague and Japan; in 1972, to Amsterdam, and the next year to Sweden. In April 1973, Nearing returned to the University of Pennsylvania, where he was named Professor Emeritus, reinstating his professorship fifty-six years after his dismissal.[59]

In 1975, Nearing published his treatise *Civilization and Beyond*, in which he reiterated his firm belief that Marxist countries offered the only workable alternative to declining Western civilization. He perceived only scant hope for what lies "beyond" civilization's inevitable fall.[60] The book garnered little interest or acclaim, and HKN later explained that the Nearings' popularity and profits stemmed from their autobiographical works, not from Nearing's polemical pieces. "The sweet books paid for the bitter ones," she claimed.[61]

Two years later, Nearing gave his final International Vegetarian Union address in Bombay, India, at the age of ninety-four. He spoke as part of a panel on "Vegetarianism from Medical, Economic and Religious Aspects." HKN also participated on a panel at the Congress entitled "Reverence for Life and Spiritual Aspects of Vegetarianism."[62]

Photographs of Nearing during this period depict his deeply lined, tanned face and a remarkably healthy affect. In 1982, he was featured as one of the "witnesses" in the Hollywood film *Reds*. In the aftermath of the film, the Nearings were written up in a "Couples" article in *People Magazine* as "chic radicals" and "America's foremost contemporary pioneers." HKN was featured prominently in the article, while Nearing, who was approaching his ninety-ninth birthday, was quoted only twice. He defined "the good life" as "one which advances man's control over nature, control over himself and his place in the universe." Nearing also characterized Ronald Reagan: "Our current President is of no greater significance than the president of a garden club." Nearing had lived through twenty American presidents "and felt a robust disdain for all of them."[63]

When HKN was asked in a 1994 interview about the Nearings' experience of old age, she responded:

> I'm the same person inside but the body is aging, the body is wearing out. Scott lived to 100, and in his last years he was not sick, he had no illnesses, no aches, no pains, no arthritis. The body was just wearing out.[64]

According to HKN, they "practiced health instead of medicine or illness."[65] However, Nearing's old age was not devoid of illness or medical care.

In the early 1970s, Dr. Mason Trowbridge asked to examine the Nearings and found that Nearing suffered from a heart arrhythmia and severe ankle edema, symptoms of pernicious anemia caused by a B12 deficiency in their vegan diet. According to Helen's personal letters, both she and Scott chose to have B12 supplements and injections during the last decades of their lives, with HKN more willing to be treated than Nearing. She wrote to Trowbridge in April 1977: "I'm taking the B12 tablets too, and when I get to the same stage in decrepitude [as Scott], you may find me a more amenable patient."[66]

The story that HKN told regarding Nearing's death was that "He wasn't sick. He didn't have a doctor, or medicine, or pain. He lived a good life, and he died a good death."[67] The Nearings shared a keen intellectual interest in death and the afterlife, reading and talking extensively about dying as Nearing aged. They disparaged hospitals and nursing homes, where strangers staved off death at great expense, and hoped to die at home in the care of family and friends.[68]

Nearing was able to do so in 1983. As his health began to fail in late 1982, Nancy Berkowitz moved into the house with the Nearings to help with his death and to be there if HKN needed to be away. "It was the three of us doing Scott's death," Berkowitz explained.[69] Nearing's health slowly deteriorated. In an August 1982 photograph taken of Nearing and Berkowitz, the previously robust Nearing stood slightly stooped and gaunt, though still smiling. The following summer, over the course of a series of gatherings for his one hundredth birthday, photographs indicated a rapid decline. In June, Nearing's extended family congregated, and from a wheelchair, he was able to address a group of young people sitting around him:

> Do the thing you believe in. Do it with all your might. And keep at it no matter what. The life we have been living is so far away from the really worthwhile goals of life that we've got to stop fooling around and move toward a new way of living.[70]

He remained a teacher to the end, as well as an apologist for purposeful living.

Hundreds of letters arrived over the course of the summer congratulating Nearing on reaching his one hundredth year. Studs Terkel wrote a letter characteristic of the sort that Nearing received, claiming, "You have been something of a beacon of light to me." Terkel described visiting Nearing a few years prior and marveling at not having been able to lift and swing the heavy axe Nearing was using to chop wood. "You have always been, stubbornly, obstinately, your own man . . . What a man!" wrote Terkel.[71]

Shortly before Nearing's actual birthday, there was a celebration with neighbors and friends at which Nearing lay in a chaise lounge draped in blankets, extremely gaunt and weak. HKN posed next to him, smiling, donning a flower wreath and appearing at ease with

Nearing's weakened state as she performed his good death story. Friends baked an oversized carrot cake with a single candle marking one century and carried in a banner that read, "The world is a better place for 100 years of Scott Nearing."[72]

In *Loving and Leaving the Good Life*, HKN described how Nearing slowly wanted only juices, then only water as "his life force was lessening."[73] Nancy Berkowitz explained, "He was this very, very frail, close to the edge, could step off any second old person that was saying, 'I'm going to stop eating.'" Berkowitz, who later became a hospice caregiver, described this choice as characteristic of a dying person who is "close to stepping over."[74] HKN wrote of Nearing's peaceful death eighteen days after his one hundredth birthday: "His body had dried up; now it was withering away, and he could tranquilly and peacefully retire from it. I was with him on his couch and quietly urged him on, the morning of August 24, 1983."[75] Nearing was cremated, and a memorial service was held on September 11 at the Blue Hill, Maine, town hall.

After Nearing's death, HKN chose to live alone, though Nancy Berkowitz and others helped maintain the house and garden. HKN struggled initially with her identity in their remaining good life project. When asked by an interviewer how she would want to be remembered, she responded, "As Scott's attachment. I enjoyed more being 'Helen and.' I don't regard myself with high esteem."[76] She wrote in a later memoir that if she were fashioning a "love" for herself, she would want someone just like Nearing, whereas the ideal woman for him would be someone like herself but "more serious, more articulate, more brains, more talent, more persistent, more spiritual, greater and beautiful in every way. That I would want for him."[77] HKN's self-deprecatory remarks were perhaps reminiscent of her Theosophical Society days when she claimed to feel incomplete without her teacher and partner; however, she had realized and publicized many remarkable, personal achievements.

In truth, Helen Knothe Nearing gradually had been assuming the leadership role at Forest Farm over the previous decade. Nearing had described his earlier retreat from politics as an extension of the Indian principle, "first a student, then a householder, then a sage."[78] Their final homestead afforded a last venue for fulfilling that principle, but HKN had become the resident good life sage at Forest Farm.

CHAPTER 6

A Good Life Alone

AS THE HOMESTEADER seeks to recreate a right relationship with the natural world, she also seeks to remake herself. Visitors still traveled to see Forest Farm, and readers continued to request books. Helen Nearing gradually recognized that "there were many things still to be done" and embraced a new role for herself as an elder woman demonstrating that she could carry on the good life work alone.[1]

Continuing to host numerous visitors proved to be one of the most challenging aspects of her work. She had claimed in an interview in Nearing's last year, "If it weren't part of our social service, we'd be happy never to talk to another stranger."[2] Though inclined to solitude, HKN remained committed to the missionary aspects of their hospitality. Her neighbor Jeanne Gaudette described HKN's hosting, writing, and speaking as "this total giving out vibration till the end." Gaudette suggested that HKN seemed "to be at peace with the giving."[3] Rebecca Gould described her hospitality in 1993:

> As she had done with thousands of visitors before [the Browns], Helen Nearing (then a very youthful 89) took them around the walled-in garden, showed them the neatly arranged compost piles, invited them into the hand-built stone house and showed off Scott Nearing's books and papers . . . Ever the modest and gracious host, Helen down-played her own efforts in light of Scott's ideals and practices. At the same time, her seemingly tireless approach to her visitors—the Browns

were only two of a number who appeared that day—betrayed her endless enthusiasm for the way of life she had chosen.[4]

HKN's boundless personal energy served her well in this capacity, though toward the end of her life, she became more "snappish and abrupt" with guests, declaring that often "it was really too much."[5] HKN continued to retreat to Florida or Mexico in the winter months to recover from the relentless stream of summertime visitors and to seek privacy for writing.

Grateful visitors and readers wrote devotional, appreciative letters to HKN describing life changes they had made or hoped to make based upon the Nearings' model. Les Oke of Norwich, Ontario, described the influence their story had on his children's lives, despite the Nearings' never having raised children on a homestead:

> Thank you for helping me find my purpose . . . What a difference you've made in our lives. I don't know of any children who are healthier or happier than ours. Their lives will be forever richer because you dared to follow your heart.[6]

Another reader claimed that "civilization had all but destroyed [his] body, soul and spirit." His doctor offered as treatment a copy of *Living the Good Life*. The grateful patient wrote to HKN, "Your book gives me renewed hope that I might still be able to climb out of this madness and find my way home to the good life."[7] In a 1993 letter, Yukio Okamura pondered with his wife how they might "live the good life" in Japan, if only he had the requisite confidence in himself and his convictions.[8] Others wrote about dreaming of the Nearings' good life, "Even if I never leave the city, in my mind I have pioneered in Vermont by reading your book." Another honest woman lamented:

> I've never had the privilege of being very anti-establishment, because I lacked the courage to move too far away from traditional responsibilities. College brought debt, and debt brought responsibility. And I had my mother to take care of. But I was with you in spirit on many occasions.[9]

This woman's recognition of her limitations, along with another letter from brothers who had tried to emulate the Nearings' lifestyle in Vermont but found themselves heavily in debt, revealed a disturbing aspect of these readers' responses.[10] They were seekers, people who heard HKN's promising rendition of a life that was good and healthy and rewarding, and they yearned for that same experience but struggled to attain it.

Romantic, pastoral visions of working the land alone were difficult to realize in the United States at late-century with inflated land prices and complex social and economic demands. Maine offered less-expensive, available land and several successful farming models. Many of the older farmers on Cape Rosier had been using organic methods for generations simply because they could not afford to use expensive chemicals. A vibrant culture of organic farmers developed in the seventies, in part due to the Nearings' example but also because of the 1971 establishment of the Maine Organic Farmers and Gardeners Association (MOFGA). MOFGA provided assistance, support, and training for organic gardeners, certified eligible organic farms, and educated consumers on the "connections between healthful food, environmentally sound farming practices, and vital local economies."[11] Though HKN would advise prospective homesteaders to secure sufficient funds and labor in order to succeed, in her public talks and writings she waxed romantic about their homesteading life. Arguably, what was most attainable in her idealized message was the promise of personal growth and perchance improved health and satisfaction through some simple, environmentally conscious choices. However, HKN rarely made these subtleties clear to her audience.[12]

Helen's niece, Cornelia Tuttle, described the significant "role change" for HKN after Nearing's death as she shifted to the center of their homesteading story.[13] She became increasingly eccentric and flamboyant, appearing wholly unselfconscious in bright, mismatched second-hand clothes or very few clothes as she and Berkowitz would strip off layers to garden. HKN sometimes would stand unabashedly in her bra talking to visitors and, with age, she became even more of a "ham" in front of a camera.[14]

HKN also was extremely generous and had a "present drawer" full of trinkets that she would give away spontaneously to close friends or new acquaintances, along with wildly colored hats and shawls that she had knitted or a newspaper article that she might have saved.[15] She maintained an almost regal dignity, particularly with her notably straight posture in old age, but would temper her royal airs with "a ninety-year-old punk hairdo." She illustrated the simplicity of caring for her short hair by claiming she could throw some bar soap on a face cloth and simply wash her hair with that.[16] Maine neighbor Gail Disney described her as a good life performer: "She was . . . almost like an actor in a way, almost like a larger-than-life performer amidst her own play. 'This is what I'm doing, this is the good life, this is how you do it.'"[17]

With her youthful vitality, HKN continued to attract young followers. Nancy Berkowitz perceived her to be a "peer," though they were forty years apart in age, and Jeanne Gaudette, a contemporary of Berkowitz's, referred to Helen as a spiritual "sister." Richard Garrett, having met the Nearings as a twenty-five-year-old back-to-the-lander in 1970, was dubbed "court photographer" by HKN and became both her confidante and a visual scribe of their good life.[18] A Maine neighbor described HKN as "happy" and "incredibly alive . . . Every cell of her body was living in nature," which gave her a vivacity that appealed to a younger audience. She also continued to provide an inspirational elder model, as a visitor to Forest Farm noted in a 1993 letter to HKN: "Thank you for being a wonderful example of how I want to age."[19]

HKN forged particularly close friendships with several talented women who proved essential to her work alone. Nancy Berkowitz, who described their relationship as being like "family," continued to offer her gardening expertise at Forest Farm.[20] She and HKN modified the garden slightly, choosing to do some things differently from

Nearing. They let one section of flowers grow wild for "fairies to go to it" and simplified their labors by buying their tomatoes and peppers.[21]

HKN developed a friendship with Ellen LaConte after Nearing's death. LaConte provided intellectual companionship in Nearing's absence and also took over the management of HKN's finances and business concerns.[22] In 1986, LaConte helped reinstate the Social Science Institute charter that had lapsed in 1981 and became its president. HKN continued as secretary and remained, nominally, in a secondary position, reluctant to claim the status that she had earned. HKN later appointed LaConte executor of her estate.[23]

HKN likewise befriended Nancy Caudle-Johnson, who helped coordinate Nearing's memorial service and who became Helen's close companion and informal archivist.[24] In 1986, HKN joined a tour that Caudle-Johnson organized to the Russian and Georgian republics of the former Soviet Union. They traveled together the next year to England and to the Findhorn community in Scotland. Caudle-Johnson encouraged HKN to work as an artist in residence at the MacDowell Colony in 1988 and the Millay Colony for the Arts in 1989, where HKN completed her final memoir. Helen also officiated at Caudle-Johnson's wedding in 1994.

Another friendship emerged with herbalist Deb Soule, founder of Avena Botanicals in Rockport, Maine. HKN had cultivated a deep interest in herbalism since the 1940s, when she met the international herbalist Juliette de Baïracli-Levy at a holistic health spa in Baja, California. HKN described having been immediately infatuated with Levy, "She was obviously an original, and she was obviously happy, and she was obviously a simple living person."[25] Soule, who shared with HKN a passion for herbs and an admiration for Levy, reconnected HKN with Levy during an herb conference in the United States in 1990.[26] Each of these friendships supported critical aspects of HKN's life and work.

However, HKN did not forge particularly close relationships with her neighbors on Cape Rosier. As in Vermont, local farmers remained distant, distrusting her communism and her hippie following.[27] An ongoing tension also existed between HKN and her seemingly more like-minded, back-to-the-land neighbors. Former neighbor Jean Hay Bright claimed, "Unfortunately [the Nearings] picked strong, bull-headed rural people just like themselves and none of us turned out to be really close followers."[28] HKN tended to be contrary toward the people on the "block"—those living along the road in close proximity to her. She criticized them for eating meat, for participating in frivolous gatherings, and for not being as strict as the Nearings in their daily lives, Eliot Coleman later explained.[29] Neighbor Jeanne Gaudette characterized HKN as "stubborn" and "ornery" and viewed her contrariness as part of the "eccentricities that we just all struggled to keep loving her through." However, the neighbors also claimed to have continued to respect HKN as "the grandmother on the block" who had "really answered the call to living a life of voluntary simplicity."[30]

In old age, HKN thus had become an increasingly complex character. She was the consummate host who would prefer never to meet another stranger, the magnanimous landowner who chose to sell property to the neighbors whom she later criticized, and a close friend and "spiritual sister" who remained oddly separate and aloof. Toward the end of her life, HKN told Gaudette that her closest friend was a stranger from Europe who "came by and bought a book": "We just hit it off, we just are soul friends. She's my soul mate."[31] HKN was at once a teacher, a fellow student, a cranky grandmother, and an elusive friend.

Along with her eccentricities, HKN maintained a highly visible public persona after Nearing's death and was invited to speak publicly at a variety of gatherings. She even introduced the Congressional candidacy of her former neighbor Jean Hay in 1993 despite her own apolitical stance.[32] Neither HKN nor Nearing ever registered to vote in Vermont or Maine, with the exception of the summer of 1986, when Helen registered to vote just once as a Democrat in Brooksville.[33] Toward the end of her life, HKN lectured at Wellesley College, Iowa State University, the University of Maine, Unity College, and the Common Ground Fair in Winslow, Maine. She addressed students at the College of the Atlantic in Bar Harbor, Maine, as late as August 13, 1995, one month before she died. HKN also received a number of awards and honors: the Adin Ballou Peace Award from the Unitarian Universalist Peace Fellowship, 1982; the Deborah Morton Award from Westbrook College, Maine, 1986; named to the Vegetarian Hall of Fame, 1991; Artist of Life Award from the International Writers Guild, 1993; and an Honorary Doctorate of Humane Letters from the University of Maine, 1994.[34]

She continued to travel extensively, returning several times to the Rio Caliente spa in Guadalajara, Mexico, where she celebrated her final birthday in February 1995. During 1985 and 1986, HKN traveled throughout Europe, participated in the aforementioned tour of the former Soviet Union, and in 1987 traveled to Switzerland, Yugoslavia, and Guatemala.[35] In 1988, at age eighty-four, HKN participated in an Elderhostel bicycle tour of Holland, and the next year returned to Holland and traveled on to Greece. She also spent a portion of her last three winters living in a small camper trailer on a friend's land in West Palm Beach, Florida.[36] Photographs from these trips confirm HKN's lifelong wanderlust and her joy in travel, as she basked in the sun at Rio Caliente, sped by on a bike in northern Holland, or admired an ancient stone wall at Delphi. The friends with whom she was photographed were strikingly younger than she. In a group photograph at Rio Caliente, HKN was the only elder, and a photograph from Greece depicted her resting in the sun using a young woman's stomach for a pillow. The images reflect her youthful spirit and remarkable vitality as she defied stereotypical notions of old age.

HKN also traveled for a third and final time to Cuba in March 1993, a trip organized by a "Let Cuba Live" group with which she had become associated. Upon her return, she gave an address at a "Let Cuba Live" event in Portland, Maine, in which she condemned the United

States' blockade of Cuba, expressing continued defiance of American foreign policy: "I've been to Cuba three times and all three times *primarily* for the same reason: the U.S. government forbids it. If they don't want me to go, I want to see what's there."[37] Such a claim epitomized HKN's politics, forged more by defiance than ideology, and hearkening back to the Nearings' former steadfast, anti–Cold War stance.

In 1992, HKN published a final memoir, *Loving and Leaving the Good Life*, in which she recounted Nearing's death and chronicled her relationships with both Nearing and Krishnamurti. This was HKN's only book attempt as a sole author in which she did not rely heavily upon photographs or other people's quotations. In the text, she resisted using the first person "I" as she heralded earlier theosophical notions of suppressing the ego in order to experience a mystical unity or whole. *Loving and Leaving the Good Life* did not present such suppression or elimination as being in deference to a patriarchal Master. Rather, HKN's choice to avoid the first person in *Loving and Leaving the Good Life* demonstrated disdain for contemporary, egocentric practices through which one loses a sense of common ground:

> What and who is this "I" anyway? We call the body "mine." We inhabit it; yet it is not us. What and who is this "I" that we aggrandize and center upon all through our lives? We are all part of the same unity-life.[38]

Her notion of the ego had evolved from a self-deprecating elimination of one's self to a recognition of the importance of one's self as part of a larger whole.[39]

In the chapter entitled "Twilight and Evening Star," HKN described Nearing's passing away. In this highly romanticized rendition, she did not mention Nancy Berkowitz's work, nor did she include any of the difficult details related to Nearing's dying process. Berkowitz described such omissions as consistent with HKN's ongoing reluctance to expose the intimate side of his death, "She never [talked] about how hard it was on her own personal heart . . . her partner of fifty

years." Berkowitz noted, however, that such details could have bene-
fited others faced with the challenges of caring for a dying spouse.
Berkowitz explained:

> There's a whole thing about Helen and Scott. And there is a myth,
> there's a mythical Helen and Scott, and there's who they really were.
> And I think people have to figure that out themselves, what part of the
> myth they're going to take on and what they can read between the lines
> and see how it really was.[40]

Loving and Leaving the Good Life exemplified the inherent tension between the mythical, romanticized good life story that HKN tena-
ciously promoted and the realities that she omitted.

As HKN approached her ninetieth birthday, she remained preoccupied with her own aging and dying, repeatedly expressing concerns
regarding the fate of her homestead. She, along with Social Science
Institute members, envisioned establishing a "Good Life Center" as
an educational, nonprofit organization that would maintain and op-
erate Forest Farm as the Nearings had. It would offer a year-round
homesteading model, a resource for those interested in the Nearings'
theories and practices, with an emphasis upon organic, sustainable
small farming, self-subsistence, and simplicity. HKN's close friend
Nancy Caudle-Johnson organized a Good Life Music Village in June
1989 as a fundraiser for the evolving Good Life Center. HKN co-
hosted the week-long music and study program on Cape Rosier with
musician Paul Winter.[41] Shortly before her death, HKN decided to
entrust the Good Life Center, including all documents and artifacts
therein, to the Trust for Public Land, a nonprofit, land conservation
organization.[42]

In the final years of her life, HKN also expressed deep concern
about who would take care of her during her dying process. She considered moving to an artists' retirement colony in Holland that she
had visited in the winter of 1992, but concluded that she would stay in
Maine. She remarked in an interview with *Down East Magazine,* "It's

crazy to flee this place [Forest Farm] because this is my skin, my life. It's got all my books and papers and music—everything I want. And I never know who's coming around the corner."[43] Friends such as Jeanne Gaudette, Nancy Berkowitz, and others assured Helen that they would care for her, and they briefly enacted a schedule whereby one woman was assigned to help her each day. HKN, however, asked that they discontinue this practice, as she needed work for visitors to do.[44]

In her final address at the College of the Atlantic given shortly before her death, HKN's tone was particularly pessimistic and dark:

> Well, you say, what about your subject? What about the Good Life? What about the artistry, the idealism, kindness, the goodness of some of mankind? The Beethovens, the DaVincis, Schweitzers, the Gandhis, the Thoreaus? A Jesus, a Buddha, a St. Francis? They, I'm sorry to say, are the few exceptions that prove the rule that most of us are not up to snuff. Most of us fritter away our time in trivialities if not in real depravity.[45]

Her later public voice became increasingly tired and foreboding, perhaps a symptom of fatigue in old age or perhaps a nearer echo of her former teacher's words.

HKN had written in her Spanish notebook in 1989 that she wanted to die by age ninety-two.[46] She likewise had described her aging in *Loving and Leaving the Good Life:*

> There was now a sense of letting go, of slowing down. The job, such as it had been, was almost done. I had been a carefree and happy traveler; now I was on the homestretch, the end was just around the bend.[47]

Her final published work, *Light on Age and Dying: Wise Words,* was a compendium of quotations regarding death and the afterlife in which she claimed, "Life has come almost full circle and I am ready to go on."[48] This last book reflected her belief in reincarnation ("I cannot think of death as more than the going out of one room into another,"

William Blake) and a curious longing for death ("No one knows whether death, which men in their fear apprehend to be the greatest evil, may not be the greatest good," Socrates.) There was a readiness about Helen Nearing.[49]

On Sunday, September 17, 1995, a particularly demanding visitor had exhausted her. HKN was especially annoyed that the visitor had brought a copy of the newly released *Light on Age and Dying* for her to autograph, and this was HKN's first time to see the book. She was furious that her publisher, Tilbury House, had sent copies to bookstores before sending one to her. She decided to seek refuge by rushing out to see the Cuban film *Strawberry and Chocolate* at the Grand Auditorium in Ellsworth, Maine. On the way there, HKN crashed her tan Subaru station wagon into an oak tree.[50] Some people speculated that she had a "mini stroke," though roads were slick with rain, and HKN—who never wore a seat belt—was a notoriously fast, reckless driver.[51]

Gaudette outlined the events after her crash: The first man to arrive on the scene of the accident assessed that HKN was badly injured but conscious. He asked her to squeeze his hand, which she did, and then he went for help. A mother and daughter stopped next, and the mother, a Reiki master, recognized HKN and immediately began performing Reiki healing.[52] Within ten minutes, an ambulance arrived and transported HKN to the Blue Hill Hospital. The doctor on duty also recognized her. He quickly procured her sister Alice Vaughn's telephone number, whom he contacted in Vermont. Vaughn, knowing that her sister would not have wanted medical intervention, requested that she not be put on artificial life support.

Three of HKN's women friends—Jeanne Gaudette, Karen Frangoulis, and Diane Fitzgerald—heard about the accident and rushed to the hospital. They stood by her in the emergency room with their hands resting on her head, quietly chanting "om." They later described having felt "all this energy rushing out of the top of [her] head," which they believed was HKN leaving her body. Her pulse stopped a few minutes later, shortly before Nancy Berkowitz arrived.[53]

Her friends brought her body back to Forest Farm, where they held a three-day vigil. People built informal altars out of memorabilia, pictures, wishing stones, candles, and apples. On September 19, Nancy Berkowitz and her husband Warren took Helen's body in a pine coffin to a crematorium. Two public memorial services were held, one at the Brooksville Town Hall and a second the next weekend at the Common Ground Fair in Windsor, Maine, where HKN was supposed to have given an address.[54]

Nancy Berkowitz reflected upon HKN's death: "It was a big thing, a big missing piece when she died. So suddenly. Even though I knew that she was ready and going in that direction, it was still a shock."[55] HKN's death coincided with the recent publication of her last book. In turn, several of her obituaries were featured alongside a review of *Light On Age and Dying* as, remarkably, Helen Nearing the consummate marketer managed posthumously to publicize her final book.[56] HKN also recently had finalized her agreement with the Trust for Public Land, which, along with her readiness to go, led to speculation regarding whether she had perhaps taken her own life. Her close friend Nancy Berkowitz dismissed the notion altogether, claiming that she would have felt suicide was bad for her karma, and that Helen Knothe Nearing would never have crashed intentionally into a tree: "She might have hurt the tree."[57]

Epilogue

AFTER HELEN KNOTHE NEARING'S DEATH, her obituary appeared in a wide variety of publications, along with articles on the Nearings' good life legacy and on *Light on Age and Dying*. The *Washington Post* published an article entitled "Two Escapees from Get-Aheadism"; the *Boston Globe* featured "Helen Nearing, Popular Author Who Praised Simple Living; at 91," and the *New York Times* ran a lengthy obituary "Helen K. Nearing, Maine Author, Dies at 91."[1] Several writers referred to HKN as a "popular author" or the Nearings as the "senior gurus of the back-to-the-land movement." In the December 31, 1995, *New York Times,* HKN was featured as one of the hundred distinguished individuals who died that year.[2] Her simple life had gained significant notoriety.

Forest Farm remained open to the public after her death; the Social Science Institute was dissolved and the Good Life Center established in its place. It would be a nonprofit 501(c)3 organization under the direction of the Trust for Public Land and a board of stewards, most of whom were affiliated with the Social Science Institute. In 1998, the Good Life Center became self-supporting with an elected president and the same board. The property presently is run by two resident stewards who maintain the house and garden, coordinate book sales, and host visitors. During the last decade, over ten thousand guests were welcomed at the Good Life Center, and Nearing texts remained in print, satisfying HKN's primary long-term goals of having books stay available and someone present to greet visitors at Forest Farm.[3]

In turn, the Nearings' good life story continues to inspire readers and guests. Forest Farm frequently is cited in travel or gardening sections of Maine periodicals and an October/November 2003 special two-hundredth issue of *Mother Earth News* highlighted the Nearings' simple-living lessons.[4] The recent popularity of the organic shopping magazine *MaryJanesFarm,* published by MaryJane Butters, epitomizes a present-day example of ongoing back-to-the-land idealism that the Nearings, like Butters, vigorously promoted.[5]

However, Helen Knothe Nearing's story also has received criticism in recent years. Books like Jean Hay Bright's memoir, *Meanwhile, Next Door to the Good Life* (2003), seek to debunk prevailing Nearing myths. Bright highlighted, in particular, contradictory claims, omissions, and misinformation regarding the Nearings' personal finances. HKN found her life's purpose in the propagation and enactment of their personal narratives, both the truths and the myths. In turn, interpretation and context remain essential for a critical understanding of her good life. As Nancy Berkowitz claimed, one must "read between the lines and see how it really was."[6]

In a 1993 interview, when asked what she would like to leave as her legacy, HKN had responded, "That they tried their best in the circumstances in which they were. They weren't perfect, but they tried to do their best."[7] Both Nearings' final mantra became a likewise humble, "Do the best that you can [in the place where you are], and be kind."[8] These modest sayings denoted fallibility and a human side to their good life, offering insight into "how it really was."[9]

Furthermore, in any deconstruction of their good life story, one finds that HKN's deep spirituality, her countercultural character, her strong sense of purpose, and her turn to the land were authentic and profound. Her vegetarianism expressed a genuine care for all living things, as well as humility and gratitude for her part in a larger ecosystem. She claimed in later life, "I'd like to be able to live on light alone," indicating abundant respect for the inherent power of natural forces.[10] She remythified rural landscapes for her readers and followers, inspiring them to make daily choices that might create simpler, more sustainable good lives in an increasingly complex world. Finally, Helen Knothe Nearing became a free-standing, energetic promoter of her good life through her daring leaps into theosophy, travel, romance, communism, homesteading, building, writing, and dying. And she thoroughly enjoyed telling the story.

Notes ✺

PROLOGUE (PP. 1–5)

1. Helen Nearing, *Loving and Leaving the Good Life* (Post Mills, Vt.: Chelsea Green Publishing, 1992), 30.

2. Jeffrey Jacob, *New Pioneers: The Back-to-the-Land Movement and the Search for a Sustainable Future* (University Park: Pennsylvania State University Press, 1997), 55–57.

3. Paul Greenberg, "The Really Simple Life: How Two Big-City Radicals Helped Take America Back to the Land," *Boston Sunday Globe,* September 5, 2004.

4. Rebecca Kneale Gould, *At Home in Nature: Modern Homesteading and Spiritual Practice in America* (Berkeley: University of California Press, 2005), 2.

5. Jacob, *New Pioneers,* 57; Greenberg, "The Really Simple Life."

6. Shari Benstock, ed., *The Private Self: Theory and Practice of Women's Autobiographical Writings* (Chapel Hill: University of North Carolina Press, 1988), 11, 15, 19.

7. Helen Nearing, *The Good Life of Helen Nearing,* interview with Tammy Simons, audiotape, 90 minutes (Sounds True Audio, 1994).

8. Ibid.

9. Helen Nearing, *Loving and Leaving the Good Life,* disclaimer.

10. Personal Narratives Group, eds., *Interpreting Women's Lives: Feminist Theory and Personal Narratives* (Bloomington: Indiana University Press, 1989), 261.

11. Helen Nearing, *Loving and Leaving the Good Life,* 75.

12. Helen Nearing, *The Good Life of Helen Nearing;* Helen and Scott Nearing, *The Good Life: Helen and Scott Nearing's Sixty Years of Self-Sufficient Living* (New York: Schocken Books, 1989), 14. The text combines in one volume *Living the Good Life* and *Continuing the Good Life.*

13. Greenberg, "The Really Simple Life."

14. Ibid.

15. Dana Goodyear, "The Simple Life, Inc.: How MaryJane Butters Reinvented the Farm Girl," *New Yorker,* October 11, 2004, 70–77.

16. Rebecca Kneale Gould, "At Home in Nature: The Religious and Cultural Work of Homesteading in Twentieth-Century America" (PhD diss., Harvard University, 1997), 313.

1. SPIRITUAL FORMATION (PP. 6–29)

1. *Helen Nearing: Conscious Living/Conscious Dying.* Produced by Polly Bennell and Andrea Savris (Bullfrog Films, 2000).
2. Cornelia Tuttle, telephone interview with Margaret Killinger, March 14, 2004.
3. Helen Nearing, *Loving and Leaving the Good Life* (Post Mills, Vt.: Chelsea Green Publishing, 1992), 24–27.
4. Helen Knothe to Scott Nearing, August 25, 1930, Nearing Collection, Thoreau Institute, Lincoln, Mass.
5. Helen Nearing, *Loving and Leaving the Good Life,* 26–27.
6. Ibid., 26–28; Nancy Caudle-Johnson, Helen Nearing's obituary leaflet, Thoreau Institute. HKN had requested that Caudle-Johnson be the author of her obituary.
7. Peter Washington, *Madame Blavatsky's Baboon: A History of the Mystics, Mediums and Misfits Who Brought Spiritualism to America* (New York: Schocken Books, 1995), 55, 69, 126.
8. Ibid., 36–37.
9. Ibid., 34, 40.
10. Ibid., 41.
11. Ibid.
12. Ibid., 12.
13. Ann Braude, *Radical Spirits: Spiritualism and Women's Rights in Nineteenth-Century America* (Boston: Beacon Press, 1989), 83.
14. Washington, *Madame Blavatsky's Baboon,* 66–68.
15. *In Memoriam: P.C. Meuleman-van Ginkel 21 Mei 1841–23* (November 1902), Thoreau Institute.
16. Jeanne Gaudette, interview with Margaret Killinger, March 3, 1999.
17. Helen Nearing, *Loving and Leaving the Good Life,* 29.
18. Ibid., 28.
19. Helen Knothe to Maria and Frank Knothe, July 2, 1923, Thoreau Institute; Helen Nearing, *Loving and Leaving the Good Life,* 34.
20. Carroll Smith-Rosenberg, *Disorderly Conduct: Visions of Gender in Victorian America* (New York: Alfred A. Knopf, 1985), 176–78.
21. Sara M. Evans, *Born for Liberty: A History of Women in America* (New York: The Free Press, 1989), 161; Helen Knothe, 1923 Diary, Thoreau Institute.
22. Helen Nearing, *Loving and Leaving the Good Life,* 34–35.
23. Washington, *Madame Blavatsky's Baboon,* 13.
24. Bruce F. Campbell, *Ancient Wisdom Revived: A History of the Theosophical Movement* (Berkeley: University of California Press, 1980), 54, 87.

25. Mary Lutyens, *Krishnamurti: His Life and Death* (New York: St. Martin's Press, 1990), 7; Helen Nearing, *Loving and Leaving the Good Life,* 34.

26. Helen Nearing, *Loving and Leaving the Good Life,* 34–35.

27. Washington, *Madame Blavatsky's Baboon,* 212.

28. Helen Knothe, diary entry, September 12, 1923, Thoreau Institute.

29. Ibid., December 4–5, 1923.

30. Lutyens, *His Life and Death,* 32.

31. Helen Nearing to Mary Lutyens, September 22, 1987, Thoreau Institute.

32. Helen Nearing, *Loving and Leaving the Good Life,* 43.

33. Frank Knothe to Helen Knothe, June 1922, Thoreau Institute.

34. Frank Knothe to Helen Knothe, December 14, 1922, Thoreau Institute.

35. Maria Knothe to Helen Knothe, December 15, 1922, Thoreau Institute.

36. Helen Knothe, diary entry, January 3, 1923, Thoreau Institute.

37. Ibid., May 11, 1923.

38. Ibid., January 2, 1923; January 3, 1923.

39. Ibid., January 26, 1923; March 19–22, 1923.

40. Claudia Clark, *Radium Girls: Women and Industrial Health Reform, 1910–1935* (Chapel Hill: University of North Carolina Press, 1997), 58.

41. Helen Knothe, diary entries, March 20; March 21; March 22, 1923, Thoreau Institute.

42. Ibid., August 20; August 21, 1923.

43. Ibid., August 19, 1923; May 10, 1923; August 20, 1923.

44. Ibid., July 23; August 16, 1923.

45. Ibid., July 20; July 11, 1923; March 27, 1923.

46. Annie Besant to Helen Knothe, February 15, 1923, Thoreau Institute.

47. Helen Nearing, *Loving and Leaving the Good Life,* 44–46.

48. Helen Knothe, diary entry, September 11, 1923, Thoreau Institute.

49. Ibid., September 13, 1923.

50. Ibid., August 20, 1923.

51. Lutyens, *His Life and Death,* 11.

52. Helen Knothe, diary entry, September 1, 1923.

53. Ibid., September 15, 1923.

54. Helen Nearing to Mary Lutyens, September 22, 1987, Thoreau Institute; Lutyens, *His Life and Death,* xvi, 48, 61; Mary Lutyens, *Krishnamurti: The Years of Awakening* (New York: Farrar, Straus, and Giroux, 1975), 207; Helen Knothe, diary entry, August 14, 1923.

55. Campbell, *Ancient Wisdom Revived,* 128.

56. Radha Rajagopal Sloss, *Lives in the Shadow with J. Krishnamurti* (Reading, Mass.: Addison-Wesley Publishing Company, 1991), 66.

57. Lutyens, *Years of Awakening*, 173.

58. Helen Knothe, diary entry, August 23, 1923, Thoreau Institute; Washington, *Madame Blavatsky's Baboon*, 218.

59. Helen Nearing to Mary Lutyens, September 22, 1987, Thoreau Institute.

60. Helen Knothe, diary entry August 15, 1923, Thoreau Institute.

61. Ibid., September 9, 1923.

62. Ibid., October 5, 1923.

63. Ibid., October 19, 1923.

64. Ibid., September 27, 1923.

65. Ibid., October 18, 1923.

66. Helen Nearing, Spanish Notebook [a journal of HKN's that happened to have "Cuaderno español" on the cover], Thoreau Institute; Helen Nearing to Mary Lutyens, September 22, 1987, Thoreau Institute.

67. Helen Knothe, diary entry, October 30, 1923, Thoreau Institute.

68. Ibid., October 31, 1923.

69. Lutyens, *His Life and Death*, 48.

70. Lady Emily Lutyens to Helen Knothe, May 9, 1924, Thoreau Institute.

71. Lady Emily Lutyens to Helen Knothe, May 13, 1924, Thoreau Institute.

72. Helen Knothe, diary entries, November 6; November 2, 1923, Thoreau Institute.

73. Ibid., November 8; November 19, 1923.

74. Ibid., December 3; December 24, 1923.

75. Ibid., December 4, 1923; November 30, 1923.

76. Ibid., December 4–5, 1923.

77. Ibid., December 7, 1923.

78. Ibid., December 31, 1923.

79. Helen Nearing, Spanish Notebook, Thoreau Institute.

80. Washington, *Madame Blavatsky's Baboon*, 213.

81. Lutyens, *His Life and Death*, 48–52; Helen Nearing, *Loving and Leaving the Good Life*, 53–54. Helen Nearing also referred to Krishnamurti's making the women cry in her Spanish Notebook, Thoreau Institute.

82. Helen Nearing, Spanish Notebook, Thoreau Institute.

83. Helen Nearing, *Loving and Leaving the Good Life*, 53.

84. Lutyens, *Years of Awakening*, 198.

85. Helen Nearing, Spanish Notebook, Thoreau Institute; Lutyens, *Years of Awakening*, 206.

86. Helen Nearing, Spanish Notebook, Thoreau Institute.

87. Lutyens, *His Life and Death,* 61; Helen Nearing, *Loving and Leaving the Good Life,* 54.

88. Lutyens, *Years of Awakening,* 236–37; Helen Nearing, *Loving and Leaving the Good Life,* 57; Helen Nearing to Mary Lutyens, September 22, 1987, Thoreau Institute.

89. Washington, *Madame Blavatsky's Baboon,* 278; Lutyens, *His Life and Death,* xiv; Campbell, *Ancient Wisdom Revived,* 129.

90. Helen Knothe, diary entry, December 4–5, 1923, Thoreau Institute.

2. POLITICAL AWAKENING (PP. 30–43)

1. Helen Nearing, Spanish Notebook, Nearing Collection Thoreau Institute, Lincoln, Massachusetts.

2. Scott and Nellie M. S. Nearing, *Women and Social Progress: A Discussion of the Biologic, Domestic, Industrial and Social Possibilities of American Women* (New York: MacMillan, 1912), x, xii, 280.

3. Scott Nearing, *The Making of a Radical: A Political Autobiography* (New York: Harper Collins, 1972), 178; Helen Nearing, *Loving and Leaving the Good Life* (Post Mills, Vt.: Chelsea Green Publishing, 1992), 21.

4. Scott Nearing, *The Making of a Radical,* 68.

5. John A. Saltmarsh, *Scott Nearing: An Intellectual Biography* (Philadelphia: Temple University Press, 1991), 108, 208, 246; Robert Nearing with Elena S. Whiteside, "Memory Keeps Me Company," personal memoir, 1998, author's collection.

6. Saltmarsh, *Scott Nearing,* v.

7. Manumit School Brochure, 1929, http://www.manumitschool.com/ManumitDocs/bro1929.htm (last accessed March 10, 2006).

8. Saltmarsh, *Scott Nearing,* 208, 246.

9. Ibid., 240.

10. Michael Gold, "Change the World!" *The Daily Worker,* January 21, 1935.

11. Ibid.; "Fellow Traveller," in *Encyclopedia of the American Left,* ed. Mari Jo Buhle, Paul Buhle, and Dan Georgakas (New York: Garland Publishing, 1990), 227; Rebecca Kneale Gould, *At Home in Nature: Modern Homesteading and Spiritual Practice in America* (Berkeley: University of California Press, 2005), 154.

12. *Helen Nearing: Conscious Living/Conscious Dying.* Produced by Polly Bennell and Andrea Sarris (Bullfrog Films, 2000).

13. Scott Nearing, *The Twilight of Empire: An Economic Interpretation of Imperialist Cycles* (New York: Vanguard Press, 1930); "Scott Nearing," in *Encyclopedia of the American Left,* 227; Saltmarsh, *Scott Nearing,* 208, 246.

14. Photographs, Thoreau Institute; Helen Nearing, *The Good Life Picture Album* (New York: Dutton/Signet, 1974).

15. Helen Nearing, *The Good Life of Helen Nearing*, interview with Tammy Simons (Sounds True Audio, 1994).

16. Robert Nearing, "Memory Keeps Me Company," 18; Helen Knothe, diary entry, October 28, 1923, Thoreau Institute.

17. Robert Nearing, "Memory Keeps Me Company," 23, 30; Robert Nearing to the Stratton Foundation, June 29, 2002, author's files.

18. Scott Nearing to Helen Knothe, November 11, 1929, Thoreau Institute.

19. Scott Nearing to Helen Knothe, letters of the 1930s dated Friday; Saturday AM; Sat 8 PM, Thoreau Institute.

20. Helen Knothe, Five Year Diary, December 4, 1936; October 25, 1938, Thoreau Institute. (This diary was a gift to Helen Knothe from her brother Alex. The cover read "Five Year Diary," and HK wrote in it from 1936 to 1940.)

21. Jerry Goldman, interview with Margaret Killinger, October 15, 2001.

22. Scott Nearing to Helen Knothe, December 20, 1929, Thoreau Institute.

23. Helen Knothe, Five Year Diary, September 9, 1937, Thoreau Institute.

24. Ibid., August 20, 1937; February 17, 1939; December 9, 1939; Helen Knothe to Scott Nearing, February, 1939, Thoreau Institute; Scott Nearing to Helen Knothe, February, 1939, Thoreau Institute; Scott Nearing to Helen Knothe, February, 1940, Thoreau Institute.

25. Helen Knothe, Five Year Diary, entry from Moscow, August 15, 1936; January 3, 1938; March 1, 1937; February 24, 1936; December 4, 1936, Thoreau Institute.

26. Saltmarsh, *Scott Nearing*, 246; Helen Nearing, *Loving and Leaving the Good Life*, 78.

27. Helen Nearing, *Loving and Leaving the Good Life*, 79; Gail Disney, interview with Margaret Killinger, March 4, 1999.

28. Helen Nearing, *Loving and Leaving the Good Life*, 70.

29. Jeff Clark, "Summing Up the Good Life," *Down East Magazine*, April 1992, 22.

30. Robert Nearing said of Helen: "She paid her dues. You lived with Scott Nearing and, damnit, you paid your dues," referring to Scott's rigidity and sternness. Presentation, Nearing Symposium, Stratton Mountain, Vermont, June 22–23, 2002.

31. Helen Nearing, *Loving and Leaving the Good Life*, 79.

32. Helen Knothe to Maria and Frank Knothe, June 18, 1930, Thoreau Institute; Helen Nearing, *Loving and Leaving the Good Life*, 70–71.

33. Scott Nearing to Helen Knothe, August 12, 1930, Thoreau Institute.

34. Ibid.

35. Helen Nearing, *Loving and Leaving the Good Life*, 79; Helen Knothe to Maria and Frank Knothe, June 18, 1930, Thoreau Institute.

36. Helen Nearing, *Loving and Leaving the Good Life*, 80, 92–93. See also Spanish Notebook, Thoreau Institute.

37. Scott Nearing to Helen Knothe, December 18, 1931, Thoreau Institute.

38. Scott Nearing to Helen Knothe, January 7, 1933, Thoreau Institute.

39. Saltmarsh, *Scott Nearing*, 2, 48–59.

40. Gould, *At Home in Nature*, 150.

41. Scott Nearing to Helen Knothe, Detroit, 1930s Letters, Thoreau Institute.

42. Scott Nearing to Helen Knothe, December 29, 1934, Thoreau Institute.

43. Scott Nearing to Helen Knothe, August 25, 1930, Thoreau Institute.

44. Helen Nearing, *Loving and Leaving the Good Life*, 68–69, 74, 80; Saltmarsh, *Scott Nearing*, 246.

45. Helen Knothe to Maria and Frank Knothe, August 23, 1931, Thoreau Institute.

46. Helen Knothe to Maria and Frank Knothe, August 23, 1931, Thoreau Institute.

47. Helen Knothe to Maria and Frank Knothe, September 4, 1932, Thoreau Institute.

48. Helen Knothe to Maria and Frank Knothe, August 23, 1931, Thoreau Institute.

49. Suian (unclear spelling, unknown identity beyond implications in letter) to Helen Knothe, August 13, 1934, Thoreau Institute.

50. Helen Knothe to Maria and Frank Knothe, February 26, 1931, Thoreau Institute.

51. Jerry Goldman, interview, October 15, 2001.

52. Helen and Scott Nearing, *The Good Life: Helen and Scott Nearing's Sixty Years of Self-Sufficient Living* (New York: Schocken Books, 1989), 127–129. The text combines in one volume *Living the Good Life* and *Continuing the Good Life*.

53. Scott Nearing to Helen Knothe, January 14, 1941, Thoreau Institute.

54. Scott Nearing to Helen Knothe, February 1, 1936, Thoreau Institute.

55. Helen Nearing, *Loving and Leaving the Good Life*; Helen Knothe to Maria and Frank Knothe, February 26, 1931, Thoreau Institute.

56. Helen Nearing, *Loving and Leaving the Good Life*, 71–72.

3. THE VERMONT EXPERIMENT (PP. 44–60)

1. Helen Knothe, Five Year Diary, Nearing Collection Thoreau Institute, Lincoln, Massachusetts.

2. "Review of Scott Nearing's *Must We Starve?*" *The New Republic*, February 22, 1933.

3. John A. Saltmarsh, *Scott Nearing: An Intellectual Biography* (Philadelphia: Temple Unversity, 1991), 245–46.

4. Vivien Ellen Rose, "Homesteading as Social Protest: Gender and Continuity in the Back-to-the-Land Movement in the United States 1890–1980 (Helen Nearing, Scott Nearing, Ralph Borsodi, Myrtle May Borsodi, Mildred Loomis)," Ph.D./SUNY Binghamton, 1997; Jeffrey Jacob, *New Pioneers: The Back-to-the-Land*

114   *Movement and the Search for a Sustainable Future* (University Park: Pennsylvania State University, 1997).

5. Rebecca Kneale Gould, "At Home in Nature: The Religious and Cultural Work of Homesteading in Twentieth-Century America," PhD dissertation, Harvard University, 1997, 313.

6. Richard Judd and Christopher S. Beach, *Natural States: The Environmental Imagination in Maine, Oregon, and the Nation* (Washington, D.C.: Resources for the Future, 2003), 2.

7. Pearl S. Buck, *The Good Earth* (New York: John Day Company, 1931); Judd and Beach, *Natural States,* 5–9; Rebecca Kneale Gould, *At Home in Nature: Modern Homesteading and Spiritual Practice in America* (Berkeley: University of California Press, 2005), 142, 163; Gould, "At Home in Nature," 313.

8. Helen Knothe wrote "grand to be alone" on the fourth of July, 1937; Five Year Diary entries, June 6, 1940; March 9, 1936; August 1, 1936; September 16, 1936; Thoreau Institute.

9. Robert Nearing, with Elena S. Whiteside, "Memory Keeps Me Company," personal memoir, 1998, author's collection, 28.

10. Gould, *At Home in Nature,* 139.

11. Saltmarsh, *Scott Nearing,* 56–57.

12. *The Nearings at Forest Farm: The Vermont Years.* Produced by Sandy Mackinnon, C. J. King, and Greg Joly, VHS, Stratton Foundation's 2002 Summer Symposium.

13. Cornelia Tuttle, telephone interview.

14. Gould, "At Home in Nature," Preface; Gould, *At Home in Nature,* 164.

15. Helen Nearing, *Loving and Leaving the Good Life* (Post Mills, Vt.: Chelsea Green Publishing, 1992), 74.

16. Jean Hay Bright, *Meanwhile, Next Door to the Good Life* (Dixmont, Maine: BrightBerry Press, 2003), 330–42. Bright described this twenty-eight-month period between December 1932 and April 1935 as "a land-buying binge" for Helen Knothe and Scott Nearing. Presentation, Nearing Symposium, Stratton Mountain, Vermont, June 22–23, 2002.

17. Helen Knothe, Five Year Diary, Thoreau Institute.

18. Scott Nearing to Helen Knothe, February 5, 1940, Thoreau Institute.

19. Richard Gregg and Helen Louise Porter Phillbrick, *Companion Plants and How to Use Them* (Old Greenwich, Conn.: Devin-Adair Company, 1966), vii.

20. *Living the Good Life* (Bullfrog Films, 1977); Helen and Scott Nearing, *The Good Life: Helen and Scott Nearing's Sixty Years of Self-Sufficient Living* (New York: Schoken Books, 1989) 31–39 (the text combines in one volume *Living the Good Life* and *Continuing the Good Life*); Helen Nearing, *Loving and Leaving the Good Life,* 104.

21. Greg Joly, presentation, Nearing Symposium, Stratton Mountain, Vermont, June 22–23, 2002; Greg Joly, "Epilogue," *The Maple Sugar Book: with Remarks on Pioneering as a Way of Life in the Twentieth Century* (White River Junction, Vt.: Chelsea Green Publishing, 2000), 259.

22. Helen and Scott Nearing, *The Maple Sugar Book,* 259–60.

23. Greg Joly, presentation, Nearing Symposium.

24. Rose, "Homesteading as Social Protest," 16–17.

25. Jerry Goldman, interview with Margaret Killinger, October 15, 2001.

26. Helen Knothe, Five Year Diary entry, May 11, 1937, Thoreau Institute.

27. Gould, *At Home in Nature,* 78, 83; Helen Nearing, Spanish Notebook, Thoreau Institute.

28. Helen Nearing, *Loving and Leaving the Good Life,* 97.

29. Scott Nearing to Helen Knothe, December 31, 1934, Thoreau Institute; Scott Nearing to Helen Knothe, January 21, 1934, Thoreau Institute.

30. Helen Knothe to Maria Knothe, 1935, Thoreau Institute.

31. Helen Knothe, Five Year Diary entries, August 20, 1937; February 17, 1939; December 9, 1939, Thoreau Institute; Helen Knothe to Scott Nearing, February, 1939, Thoreau Institute; Scott Nearing to Helen Knothe, February, 1939, Thoreau Institute; Scott Nearing to Helen Knothe, February, 1940, Thoreau Institute.

32. Helen Knothe to Scott Nearing, January 1944, Thoreau Institute.

33. Scott Nearing to Helen Knothe, January 18, 1944, Thoreau Institute.

34. Kristin McMurran, "A Jug of Carrot Juice, A Loaf of Bread and One Another in the Wilderness Give the Nearings Paradise Now," *People Magazine,* August 23, 1982, 78.

35. In HK's collection of letters, there were two letters of condolences and concern after her surgery: Scott Nearing to Helen Knothe, January 1944; Krishnamurti to Helen Knothe, February 11, 1944, Thoreau Institute.

36. Helen Knothe, automatic writing, April 12, 1944, Thoreau Institute.

37. Helen Nearing, *Spanish Notebook,* Thoreau Institute.

38. Scott Nearing to Helen Knothe, November 11, 1931, Thoreau Institute.

39. Frank Lloyd Wright, quoted by Helen and Scott Nearing, in *The Good Life: Helen and Scott Nearing's Sixty Years of Self-Sufficient Living,* 85–86, 65–67.

40. "Some History of the Nearing Stone Houses in Vermont," handout, Nearing Symposium, Stratton Mountain, Vermont, June 22–23, 2002.

41. Helen Nearing, *The Good Life of Helen Nearing,* audiocassette (Sounds True Audio, 1994).

42. Scott Nearing to Helen Knothe, 1940s letters, Thoreau Institute.

43. Helen and Scott Nearing, *The Good Life: Helen and Scott Nearing's Sixty Years of*

Self-Sufficient Living, 17; Gail Disney, interview with Margaret Killinger, May 17, 2006.

44. Jerry Goldman, interview.

45. Gene Lepkoff, interview with Margaret Killinger, tape recording, December 29, 2001.

46. Greg Joly, "Epilogue," *Maple Sugar Book,* 261.

47. Jerry Goldman, interview.

48. Helen Knothe, Five Year Diary entries, December 25, 1936; March 26, 1937, Thoreau Institute.

49. Helen and Scott Nearing, *The Good Life: Helen and Scott Nearing's Sixty Years of Self-Sufficient Living,* 182–87; *Boston Daily Globe,* March 5, 1945.

50. Jerry Goldman, interview.

51. Anne Wilkes Tucker, Claire Cass, and Stephen Daiter, ed., *This Was the Photo League: Compassion and the Camera from the Depression to the Cold War* (Chicago: Stephen Daiter Gallery, 2001).

52. Collamer M. Abbott, "They Get Away From It All in Pikes Falls," *The Vermont Phoenix,* November 12, 1948.

53. Jerry Goldman, interview; Gene Lepkoff, interview; John Saltmarsh, talk at Forest Farm Monday Night Meetings, Harborside, Maine, July 2000.

54. Robert Ellsberg, ed., *By Little and By Little, The Selected Writings of Dorothy Day* (New York: Alfred A. Knopf, 1983).

55. Jerry Goldman, interview; Rebecca Lepkoff, interview.

56. Robert Nearing, *Memory Keeps Me Company,* 119–20.

57. Helen Nearing, *Loving and Leaving the Good Life,* 82.

58. Helen Nearing to Maria Knothe, December 14, 1947, Thoreau Institute.

59. Jeff Clark, "Summing Up the Good Life," *Down East Magazine,* April 1992, 22.

60. Helen and Scott Nearing, *The Maple Sugar Book,* xi.

61. "Maple Sugaring Past and Present," *Agriview,* March 1, 1950, Nearing Collection, Howard Gottlieb Archival Research Center, Boston University, Boston, Massachusetts.

62. Rose, "Homesteading as Social Protest," 21.

63. Rebecca Kneale Gould, presentation, Nearing Symposium, Stratton Mountain, Vermont, June 22–23, 2002.

64. Invitation by Paul S. Erikson to Sugaring-Off Party at John Day Company Publishers, Thoreau Institute; Greg Joly, "Epilogue," *The Maple Sugar Book,* 255–57; Greg Joly, presentation, Nearing Symposium.

65. Aunt Tabitha, "Ravelings," *The Brattleboro Reformer,* March 2, 1950.

66. Helen Nearing, *The Good Life Picture Album* (New York: Dutton/Signet, 1974), 65.

67. Nearing Skinner, "Vermonters Reflect on Nearing Legacy," *Rutland* [Vermont] *Herald,* September 1995.

68. Scott Nearing to Helen Knothe, December 1928, Thoreau Institute.

69. Jeanne Gaudette, interview with Margaret Killinger, March 3, 1999.

4. LIVING THE GOOD LIFE (PP. 61–73)

1. Studs Terkel, *American Dreams: Lost and Found* (New York: Pantheon Books, 1980), 326.

2. Jerry Goldman, interview with Margaret Killinger, October 15, 2001; Rebecca Lepkoff, interview with Margaret Killinger, December 29, 2001.

3. Greg Joly, *A Love Greater than 70 Bushels of Baked Potatoes: Helen and Scott Nearing in Vermont.* A One Act Play. Performed at the Stratton Mountain Summer Symposium, June 23, 2002.

4. David L. Graham, "The Ruggedest, Individualest Man in Maine," *The Maine Times,* June 6, 1969, 8–9.

5. No mortgage was reported with this sale. Jean Hay Bright, *Meanwhile, Next Door to the Good Life* (Dixmont, Maine: Brightberry Press, 2003), 342.

6. Richard Judd and Christopher S. Beach, *Natural States: The Environmental Imagination in Maine, Oregon, and the Nation* (Washington, D.C.: Resources for the Future, 2003), 4–10.

7. Scott Nearing, *The Making of a Radical: A Political Autobiography* (New York: Harper Collins, 1972), 165, 171; John Saltmarsh, talk at Forest Farm Monday Night Meetings, Harborside, Maine, July 2000; Scott Nearing, *Man's Search for the Good Life* (Harborside, Maine: Social Science Institute, 1954), vi.

8. Helen Nearing to Maria Knothe, October 30, 1952, Nearing Collection, Thoreau Institute, Lincoln, Massachusetts.

9. Helen Nearing to Maria Knothe, October 1952, Thoreau Institute.

10. Maria Knothe to Helen Nearing, February 23, 1953, Thoreau Institute.

11. The texts were published by the Social Science Institute and could be boxed together and purchased for $5.00 or bought separately, $2.50 for *Man's Search for the Good Life* and $3.50 for *Living the Good Life,* at a small number of bookstores, through Leftist periodicals such as *Doubt* and *Interpreter,* or by mail order from the Social Science Institute in Harborside. *The Rural Mailbox,* June 1955, Nearing Collection, Howard Gottlieb Archival Research Center, Boston University, Boston, Massachusetts.

12. Scott Nearing, *Man's Search for the Good Life,* 22.

13. Ibid., 102, 90, 117.

14. Helen and Scott Nearing, *The Good Life: Helen and Scott Nearing's Sixty Years of*

Self-Sufficient Living (New York: Schocken Books, 1989), 7. The text combines in one volume *Living the Good Life* and *Continuing the Good Life.*

15. Ibid., 3, 5, 7.

16. Ibid., 65, 8.

17. Alan Trachtenberg, *Reading American Photographs: Images as History Matthew Brady to Walker Evans* (New York: Hill and Wang, 1989), 164, 192, 246; Helen Nearing, *The Good Life Album of Helen and Scott Nearing* (New York: Dutton/ Signet, 1974), dedication.

18. Helen and Scott Nearing, *The Good Life: Helen and Scott Nearing's Sixty Years of Self-Sufficient Living,* 130.

19. Rebecca Kneale Gould, *At Home in Nature: Modern Homesteading and Spiritual Practice in America* (Berkeley: University of California Press, 2005), 3; Rebecca Kneale Gould, presentation, Nearing Symposium, Stratton Mountain, Vermont, June 22–23, 2002.

20. Gould, *At Home in Nature,* 3; John Saltmarsh, *Scott Nearing: An Intellectual Biography* (Philadelphia: Temple University Press, 1991), 254; Rebecca Gould, presentation, Nearing Symposium; Gould, handout, talk at Forest Farm Monday Night Meetings, Harborside, Maine, July, 2000.

21. Judd and Beach, *Natural States,* 4.

22. "Maple Syrup Time," http:home.earthlink.net/~jimcapaldi/maplesyrup.htm (accessed February 2004).

23. Helen and Scott Nearing, *USA Today: Reporting Extensive Journeys and First-hand Observations Commenting on Their Meaning and Offering Conclusions Regarding Present-Day Trends in the Domestic and Foreign Affairs of the United States* (Harborside, Maine: Social Science Institute, 1955), xxiv, 241.

24. Helen and Scott Nearing, *The Brave New World* (Harborside, Maine: Social Science Institute, 1958), 1–4, 7, 244.

25. David E. Shi, *The Simple Life: Plain Living and High Thinking in American Culture* (New York: Oxford University Press, 1985), 257; *Encyclopedia of the American Left,* ed. Mari Jo Buhle, Paul Buhle, and Dan Georgakas (New York: Garland Publishing, 1990), 516; History of the International Vegetarian Union, http://www.ivu .org/history/ (accessed January 22, 2004).

26. Yuan-Li Wu, *Tiananmen to Tiananmen: China Under Communism 1947–1996, After Delusion and Disillusionment a Nation at a Crossroads* (College Park: School of Law, University of Maryland, 1997), 49.

27. Norman L. Rosenberg and Emily S. Rosenberg, *In Our Times: America since World War II,* fifth ed. (Englewood Cliffs, N.J.: Prentice Hall, 1995), 10–17.

28. Helen and Scott Nearing, *Socialists Around the World* (New York: Monthly Review Press, 1958), 159–60.

29. "The Socialist Century," *Daily Chronicle,* Georgetown, Trinidad, January 20, 1961, Nearing Collection Thoreau Institute, Lincoln, Massachusetts; Scott Nearing to Helen Nearing, March 14, 1963, Thoreau Institute; Helen Nearing, "Let Cuba Live" speech, audiocassette, April 16, 1993, The Good Life Center.

30. Note on photograph marked "Delhi November '67," Thoreau Institute.

31. Helen and Scott Nearing, *USA Today,* xxi.

32. Scott Nearing to Helen Nearing, January 18, 1961, Thoreau Institute.

33. Rebecca Kneale Gould, "At Home in Nature: The Religious and Cultural Work of Homesteading in Twentieth-Century America," PhD dissertation, Harvard University, 1997, 301.

34. Scott Nearing to Helen Nearing, January 21, 1952, Thoreau Institute.

35. Graham, "The Ruggedest, Individualest Man," 8–9.

36. Photographs, Thoreau Institute.

37. Helen Nearing, *The Good Life Picture Album* (New York: Dutton/Signet, 1974), 13.

38. Helen Nearing, *The Good Life of Helen Nearing,* interview with Tammy Simons, audiotape, 90 minutes (Sounds True Audio, 1994); Martha Gottlieb, interview with James Moreira, September 22, 2001, transcript, MOGFA Oral History Project, Northeast Archives, Orono, Maine; Helen Nearing, *The Good Life Picture Album,* 77, 11.

39. Helen Nearing, Spanish Notebook, Thoreau Institute; Scott Nearing to Helen Nearing, January 21, 1952, Thoreau Institute; *Juliette of the Herbs,* VHS, Mabinogion Films production, directed by Tish Streeten (Wellspring Media, 1998).

5. BACK TO THE LAND (PP. 74–91)

1. Richard Judd, lecture in environmental history course, University of Maine, April 26, 2000.

2. Samuel P. Hays, *Beauty, Health and Permanence: Environmental Politics in the United States, 1955–1985* (Cambridge: Cambridge University Press, 1987), chapter 1.

3. Rachel Carson, *Silent Spring* (Boston: Houghton Mifflin, 1962). Ralph Lutts, in his article "Chemical Fallout: Rachel Carson's *Silent Spring,* Radioactive Fallout, and the Environmental Movement," makes a compelling argument for Carson's having launched the American environmental movement. *Environmental Review* 9 (Fall 1985): 211–25.

4. Helen and Scott Nearing, *The Good Life: Helen and Scott Nearing's Sixty Years of*

Self-Sufficient Living (New York: Schocken Books, 1989), 130. The text combines in one volume *Living the Good Life* and *Continuing the Good Life.*

5. Richard Judd and Christopher S. Beach, *Natural States: The Environmental Imagination in Maine, Oregon and the Nation* (Washington, D.C.: Resources for the Future, 2003), 20.

6. Hays, *Beauty, Health, and Permanence,* 13.

7. Kirkpatrick Sale, *The Green Revolution: The American Environmental Movement, 1962–1992* (New York: Hill and Wang, 1993), Timeline, 19.

8. Jeffrey Jacob, *New Pioneers: The Back-to-the-Land Movement and the Search for a Sustainable Future* (University Park: Pennsylvania State University, 1997), 231.

9. Judd and Beach, *Natural States,* 240; David E. Shi, *In Search of the Simple Life: American Voices, Past and Present* (Salt Lake City: Peregrine Smith Books, 1986), 297. William C. Kelly, "Rodale Press and Organic Gardening," *HortTechnology* (April/
June 1992): 270.

10. Barry Commoner, *The Poverty of Power: Energy and the Economic Crisis* (London: Jonathon Cape, 1976); Wendell Berry, *The Unsettling of America: Culture and Agriculture* (San Francisco: Sierra Club Books, 1977), 7, 193.

11. Judd and Beach, *Natural States,* 240.

12. Rebecca Kneale Gould, "At Home in Nature: The Religious and Cultural Work of Homesteading in Twentieth-Century America, PhD dissertation, Harvard University, 1997, 312.

13. Harold Henderson, *The Reader* (Chicago, Illinois), June 8, 1979, Nearing Collection, Howard Gottlieb Archival Research Center, Boston University, Boston, Massachusetts.

14. Aldo Leopold, *A Sand County Almanac: With Other Essays on Conservation from Round River* (New York: Ballantine Books, 1970); Richard Gregg and Helen Louise Porter Phillbrick, *Companion Plants and How to Use Them* (Old Greenwich, Conn.: Devon-Adair Company, 1966).

15. Charles Elliot, "Up On the Farm," *Time Magazine,* January 18, 1971.

16. Jim Hupp, review of *Living the Good Life, Freedom News,* November 1970; review of *Living the Good Life, The Maple Sugar Book,* and *The Making of a Radical, Catholic Worker,* June 1972, 4; "The Nation Congratulates," *Nation,* November 16, 1970, 485; "Three Books by the Nearings," *Natural Life Styles* 1 (1971): 66.

17. John Saltmarsh, *Scott Nearing: An Intellectual Biography* (Philadelphia: Temple University Press, 1991), 263; Shi, *In Search of the Simple Life,* 257.

18. Saltmarsh, *Scott Nearing,* 263.

19. Helen Nearing, "Summing Up," *Down East Magazine* 35 (July 1979): 60–61.

20. Shi, *In Search of the Simple Life,* 256–57.

21. Rebecca Kneale Gould, *At Home in Nature: Modern Homesteading and Spiritual Practice in America* (Berkeley: University of California Press, 2005), 54.

22. Nancy Berkowitz, interview with Margaret Killinger, April 2, 2001.

23. Gail Disney, interview with Margaret Killinger, March 5, 1999.

24. Helen Nearing, "Summing Up," 62.

25. Jean Hay, interview with James Moreira, September 23, 2000, MOGFA Oral History Project Northeast Archives, Orono, Maine; Nancy Berkowitz, interview, 2001.

26 Nancy Berkowitz, interview with Margaret Killinger, tape recording, March 3, 1999. Nancy Richardson changed her name to Berkowitz when she married Warren Berkowitz in 1984. To avoid confusion, she will be referred to throughout as Nancy Berkowitz.

27. Helen Nearing, "Summing Up," 63.

28. Ibid., 61; Berkowitz, interview, 2001.

29. Helen Nearing, "Summing Up," 62.

30. Richard Garrett, "What the Nearings Have Meant to Us," *Country Journal,* 1978, 69.

31. Richard Garrett, interview with Margaret Killinger, May 17, 2006.

32. Helen Nearing, "Summing Up," 60.

33. Ibid., 60–61; Jean Hay, interview; Gail Disney, interview; Maine Organic Farmers and Growers Association (MOFGA) web side, http://www.mofga.org (accessed March, 2004).

34. David E. Shi, *The Simple Life: Plain Living and High Thinking in American Culture* (New York: Oxford University Press, 1985), 257–58.

35. Helen Nearing, Spanish Notebook, Nearing Collection Thoreau Institute, Lincoln, Massachusetts.

36. Richard Kyle, *The New Age Movement in American Culture* (Lanham, Md.: University Press of America, 1995). Washington, *Madame Blavatsky's Baboon: A History of the Mystics, Mediums and Misfits Who Brought Spiritualism to America* (New York: Schocken Books, 1995), 361.

37. Helen Nearing, *The Good Life Picture Album* (New York: Dutton/Signet, 1974); Helen Nearing, *Wise Words on the Good Life* (New York: Schocken Books, 1980); Scott Nearing, *The Making of a Radical: A Political Autobiography* (New York: Harper Collins, 1972); Scott Nearing, *Civilization and Beyond: Learning from History* (Harborside, Maine: Social Science Institute, 1975); Scott Nearing, *Free Born: An Unpublishable Novel* (Freeport, N.Y.: Books for Libraries Press, 1972, first published 1932); Helen and Scott Nearing, *Building and Using Our Sun-Heated Greenhouse* (Charlotte, Vt.: Garden Way Publishing, 1977); and Helen and Scott Near-

ing, *Continuing the Good Life: Half a Century of Homesteading* (New York: Schocken Books, 1979).

38. Helen Nearing, *The Good Life Picture Album,* 126.

39. Gail Disney, interview.

40. *Living the Good Life* (Bullfrog Films, 1977).

41. Helen Nearing, "Summing Up," 63.

42. Nancy Berkowitz, interview, 2001; Helen and Scott Nearing, *Continuing the Good Life,* 321; Helen Nearing, "Introduction," in *Our Home Made of Stone: Building in Our Seventies and Nineties* (Camden, Maine: Downeast Books, 1983).

43. Noel Perrin, "Scything the Meadow at 95," *New York Times Book Review,* May 6, 1979.

44. Kathy Gunst, "Ripping Apart the American Style of Eating," *The Maine Times,* May 15, 1987, 27.

45. Joe Allen, "Happenings," *Ad Vantages,* Deer Isle, Maine, February 1981.

46. Helen Nearing claimed to have received an inheritance that funded the project. Recent royalties from books and profits from land sales also would have contributed to the development of the property. Helen Nearing, "Introduction," *Our Home Made of Stone.*

47. Ibid.

48. Helen Nearing, *Loving and Leaving the Good Life* (Post Mills, Vt.: Chelsea Green Publishing, 1992), 176.

49. Jeanne Gaudette, interview with Margaret Killinger, March 3, 1999.

50. Nancy Berkowitz, interview, 1999.

51. Nancy Berkowitz, interview, 2001.

52. Helen Nearing, *Our Home Made of Stone,* iii.

53. Norm Lee, "Romancing the Stone," *East West Journal,* October 1984, 80–82.

54. Nancy Berkowitz, interview, 2001.

55. Nancy Berkowitz, interview, 2001; "Scott Nearing," *Encyclopedia of the American Left,* ed. Mari Jo Buhle, Paul Buhle, and Dan Georgakas (New York: Garland Publishing, 1990), 52.

56. Jane M. Hamel to Helen Nearing, May 19, 1993, Thoreau Institute. Excerpt courtesy of notes by Mary Beausoleil.

57. Jean Hay Bright, *Meanwhile, Next Door to the Good Life* (Dixmont, Maine: BrightBerry Press, 2003), 343.

58. Betta Stothart, "A Tale of Two Lives. The Nearings' Farm Gets a Fresh Coat of Paint in a New Book," *The Maine Times,* June 28, 1991. Stanley Joseph later wrote *The Maine Farm: A Year of Country Life* (New York: Random House, 1991), with photographs by his wife Lynn Karlin illustrating their flower gardens and life at the Nearings' former homestead.

59. Professor Emeritus certificate, Good Life Center, Harborside, Maine.

60. Franklin Zahn, review of *Civilization and Beyond, Win Magazine,* June 8, 1976, 20.

61. Jerry Buckley, "Living the Good Life in New England," *Newsweek,* August 29, 1983.

62. Photographs and program from the International Vegetarian Union Congress, December 7–10, 1977, Thoreau Institute.

63. Kristin McMurran, "A Jug of Carrot Juice, A Loaf of Bread and One Another in the Wilderness Give the Nearings Paradise Now," *People Magazine,* August 23, 1982, 77–79; Jerry Buckley, "Living the Good Life in New England," *Newsweek,* August 29, 1983.

64. Helen Nearing, *The Good Life of Helen Nearing,* interview with Tammy Simons (Sounds True Audio, 1994).

65. Helen Nearing, commencement address notes, College of the Atlantic, June 2, 1984, Thoreau Institute.

66. Helen Nearing to Mason Trowbridge, April 7, 1977, Thoreau Institute; Helen Nearing, *The Good Life of Helen Nearing.*

67. "Carrying on the Good Life," *Down East Magazine,* November 1985, 42.

68. Helen Nearing, *Loving and Leaving the Good Life,* 166–67.

69. Nancy Berkowitz, interview, 2001.

70. James P. Brown, "Scott Nearing: 100 Years of 'The Good Life,'" *Down East Magazine* 30 (August 1983): 171.

71. Studs Terkel to Scott Nearing, August 1983, Helen Nearing Scrapbook, Good Life Center.

72. Photographs, Thoreau Institute.

73. Helen Nearing, *Loving and Leaving the Good Life,* 183.

74. Nancy Berkowitz, interview, 2001.

75. Helen Nearing, *Loving and Leaving the Good Life,* 184.

76. Vivien Ellen Rose, "Homesteading as Social Protest: Gender and Continuity in the Back-to-the-Land Movement in the United States 1890–1980 (Helen Nearing, Scott Nearing, Ralph Borsodi, Myrtle May Borsodi, Mildred Loomis)," PhD dissertation, SUNY Binghamton, 1997, 19.

77. Helen Nearing, *Loving and Leaving the Good Life,* 82.

78. The Plowboy Interview, *Mother Earth News,* 1971, 6.

6. A GOOD LIFE ALONE (PP. 92–103)

1. Rebecca Kneale Gould, *At Home in Nature: Modern Homesteading and Spiritual Practice in America* (Berkeley: University of California Press, 2005), 230. Helen Nearing, *Loving and Leaving the Good Life* (Post Mills, Vt.: Chelsea Green Publishing, 1992), 2.

2. Kristin McMurran, "A Jug of Carrot Juice, A Loaf of Bread and One Another in the Wilderness Give the Nearings Paradise Now," *People Magazine,* August 23, 1982, 79.

3. Jeanne Gaudette, interview with Margaret Killinger, March 3, 1999.

4. Rebecca Kneale Gould, "At Home in Nature: The Religious and Cultural Work of Homesteading in Twentieth-Century America," PhD dissertation, Harvard University, 1997, 2.

5. Ellen LaConte, *On Light Alone: A Guru Meditation on the Good Death of Helen Nearing* (Stockton Springs, Maine: Looseleaf Press, 1996), 89–92.

6. Les Oke to Helen Nearing, January 28, 1993, Nearing Collection, Thoreau Institute, Lincoln, Massachusetts. Excerpt courtesy of notes by Mary Beausoleil.

7. James Woodman to Helen Nearing, August 12, 1993, Thoreau Institute. Excerpt courtesy of notes by Mary Beausoleil.

8. Yukio Okamura to Helen Nearing, September 9, 1993, Thoreau Institute. Excerpt courtesy of notes by Mary Beausoleil.

9. Kathryn Kester to Helen Nearing, October 22, 1992, Thoreau Institute; Linda Crowley Horger to Helen Nearing, August 27, 1993, Thoreau Institute. Excerpts courtesy of notes by Mary Beausoleil.

10. Steve Weiss to Helen Nearing, April 12, 1992, Thoreau Institute. Excerpt courtesy of notes by Mary Beausoleil.

11. Maine Organic Farmers and Growers Association (MOFGA) web site, http://www.mofga.org (accessed March 2004).

12. Jean Hay, interview with James Moreira, September 23, 2000, MOGFA Oral History Project, Northeast Archives, Orono, Maine.

13. Cornelia Tuttle, telephone interview with Margaret Killinger, March 14, 2004.

14. Nancy Berkowitz, interview with Margaret Killinger, April 2, 2001; Jeanne Gaudette, interview.

15. Nancy Berkowitz, interview, 2001.

16. Nancy Berkowitz, interview with Margaret Killinger, March 3, 1999; Jeanne Gaudette, interview.

17. Gail Disney, interview with Margaret Killinger, March 5, 1999.

18. Richard Garrett, interview with Margaret Killinger, May 17, 2006.

19. Jeanne Gaudette, interview; Nancy Berkowitz, interview, 2001; Gail Disney, interview; Terry Jones to Helen Nearing, July 6, 1993, Thoreau Institute. Excerpt courtesy of notes by Mary Beausoleil.

20. Nancy Berkowitz, interview, 2001; Jeanne Gaudette, interview. Nancy Berkowitz, interview, 2001.

21. Nancy Berkowitz, interview, 2001.

22. Ibid.

23. Helen Nearing, November 11, 1990, Spanish Notebook, Thoreau Institute; Nancy Caudle-Johnson, interview with Margaret Killinger, January 24, 2004.

24. Nancy Caudle changed her name to Caudle-Johnson when she married Douglas N. Johnson in 1994. To avoid confusion, she will be referred to as Nancy Caudle-Johnson.

25. Helen Nearing, *Juliette of the Herbs*, Mabinogion Films production, directed by Tish Streeten (Wellspring Media, 1998).

26. Deb Soule, interview with Pauleena Macdougall, September 23, 2000, MOGFA Oral History Project, Northeast Archives, Orono, Maine.

27. Richard Garrett, interview; Gail Disney, interview.

28. Jean Hay, interview.

29. Murray Carpenter, "The Riddle of Self-Sufficiency," *The Maine Times*, August 12, 1999, 6.

30. Jeanne Gaudette, interview.

31. Ibid.

32. Helen Nearing, loose papers, Thoreau Institute; Jean Hay Bright, *Meanwhile, Next Door to the Good Life* (Dixmont, Maine: Bright Berry Press, 2003), 298.

33. Town Offices of Brooksville, Maine, and Winhall, Vermont.

34. Helen Nearing's obituary pamphlet, Thoreau Institute.

35. Photographs, Thoreau Institute; Helen Nearing, Travel Notebook 1985–1986, Thoreau Institute.

36. Nancy Berkowitz, interview, 1999.

37. Helen Nearing, "Cuba Talk," April 16, 1993, Good Life Center, audiocassette.

38. Helen Nearing, *Loving and Leaving the Good Life*, 6.

39. Ibid., 38; Jeanne Gaudette, interview.

40. Nancy Berkowitz, interview, 1999.

41. Program for the Good Life Music Village, Thoreau Institute.

42. Peter Forbes of the Trust for Public Land, presentation, Nearing Symposium, Stratton, Vermont, June 22–23, 2002. Trust for Public Land web site, http://www.tpl.org (accessed March 2, 2004).

43. Jeff Clark, "Summing Up the Good Life," *Down East Magazine*, April 1992, 22.

44. Jeanne Gaudette, interview.

45. Helen Nearing, notes for address at College of the Atlantic, August 13, 1995, Thoreau Institute.

46. Helen Nearing, *The Good Life of Helen Nearing*, interview with Tammy Simons (Sounds True Audio, 1994); Spanish Notebook, Thoreau Institute.

47. Helen Nearing, *Loving and Leaving the Good Life*, 191.

126

48. Helen Nearing, *Light on Age and Dying: Wise Words* (Gardiner, Maine: Tilbury House, 1995), x.

49. Gaudette, interview. Helen Nearing also had been extremely upset after Stanley Joseph took his own life in April 1995. Joseph had suffered from bouts of acute depression and committed suicide by carbon monoxide asphyxiation from his running car on his property, the Nearings' former homestead. Bright, *Meanwhile, Next Door to the Good Life,* 124.

50. Nancy Caudle-Johnson, interview; LaConte, *On Light Alone,* 92.

51. Jeanne Gaudette, interview; Gail Disney, interview; Nancy Caudle-Johnson, interview.

52. Reiki healing is an ancient practice of applying Chi—the Chinese or martial arts term for an underlying force in the universe—to strengthen another's energy source or aura, usually by gentle closeness of the healer's hand. Diane Stein, *Essential Reiki: A Complete Guide to an Ancient Healing Art* (Freedom, Calif.: Crossing Press, 1995).

53. Jeanne Gaudette, interview.

54. LaConte, *On Light Alone,* 92.

55. Nancy Berkowitz, interview, 1999.

56. Betta Stothart, "Helen Nearing Parting Gifts, a Book and a Future at Forest Farm," *The Maine Times,* September 21, 1995.

57. Nancy Berkowitz quoted in *Helen Nearing: Conscious Living/Conscious Dying.* Produced by Polly Bennell and Andrea Sarris (Bullfrog Films, 2000).

EPILOGUE (PP. 104–106)

1. Colman McCarthy, "Two Escapees from Get-Aheadism," *Washington Post,* September 23, 1995; Pamela M. Walsh, "Helen Nearing, Popular Author Who Praised Simple Living; at 91," *Boston Globe,* September 18, 1995; John McQuiston, "Helen K. Nearing, Maine Author, Dies at 91," *New York Times,* September 19, 1994, B8.

2. McQuiston, "Helen K. Nearing," B8; Anne Raver, "Helen K. Nearing: Greener, Saner, Simpler," *New York Times,* December 31, 1995, 35.

3. Fundraising letter from the Good Life Center, November 2003, author's files; Nancy Berkowitz, interview with Margaret Killinger, April 2, 2001.

4. Linnea Johnson, "Nearing Enough," *Mother Earth News,* October/November 2003, 22–27; Clarke Canfield, "Nearing Spirit Lives On," *Bangor Daily News,* September 22–23, 2000, Section G, 1–2.

5. *MaryJanesFarm: Simple Solutions for Everyday Organic,* http://www.maryjanesfarm.org (accessed March 5, 2004).

6. Nancy Berkowitz, interview with Margaret Killinger, March 3, 1999.

7. Helen Nearing, *The Good Life of Helen Nearing,* interview with Tammy Simons (Sounds True Audio, 1994). 127
8. Elka Schumann, "In Memory of Scott Nearing," *The Chronicle,* August 31, 1983, Helen Nearing scrapbook, Good Life Center.
9. Nancy Berkowitz, interview, 1999.
10. Rebecca Kneale Gould, *At Home in Nature: Modern Homesteading and Spiritual Practice in America* (Berkeley: University of California Press, 2005), 88.

Bibliography ❧

MANUSCRIPT COLLECTIONS

Nearing Papers. The Good Life Center, Harborside, Maine.

Nearing Collection. The Howard Gottlieb Archival Research Center, Boston University, Boston, Massachusetts.

Nearing Collection. The Thoreau Institute, Lincoln, Massachusetts.

ORAL INTERVIEWS

Berkowitz, Nancy, interview with Margaret Killinger, March 3, 1999, tape recording.

Berkowitz, Nancy, interview with Margaret Killinger, April 2, 2001, tape recording.

Berkowitz, Nancy, telephone interview with Margaret Killinger, March 20, 2004.

Caudle-Johnson, Nancy, interview with Margaret Killinger, January 24, 2004, tape recording.

Disney, Gail, interview with Margaret Killinger, March 5, 1999, tape recording.

Garrett, Richard, interview with Margaret Killinger, May 17, 2006.

Gaudette, Jeanne, interview with Margaret Killinger, March 3, 1999, tape recording.

Goldman, Jerry, interview with Margaret Killinger, October 15, 2001, tape recording.

Gottlieb, Martha, interview with James Moreira, September 22, 2001. MOFGA Oral History Project, Northeast Archives, Orono, Maine, transcript.

Hay, Jean, interview with James Moreira, September 23, 2000. MOFGA Oral History Project, Northeast Archives, Orono, Maine, transcript.

Lepkoff, Rebecca, and Gene Lepkoff, interview with Margaret Killinger, December 29, 2001, tape recording.

Soule, Deb, interview with Pauleena Macdougall, September 23, 2000. MOFGA Oral History Project, Northeast Archives, Orono, Maine, transcript.

Tuttle, Cornelia, telephone interview with Margaret Killinger, March 14, 2004.

York, Chaitanya, interview with Anu Dudley, September 21, 2001. MOFGA Oral History Project, Northeast Archives, Orono, Maine, transcript.

The Helen and Scott Nearing Symposium. Stratton Mountain, Vermont, June 22–23, 2002.

NEARING TEXTS

Nearing, Helen. *The Good Life of Helen Nearing*. Interview with Tammy Simons. Audiocassette, 90 minutes. Sounds True Audio, 1994.

———. *The Good Life Picture Album.* New York: Dutton/Signet, 1974.

———. *Light on Age and Dying: Wise Words.* Gardiner, Maine: Tilbury House, 1995.

———. *Loving and Leaving the Good Life.* Post Mills, Vt.: Chelsea Green Publishing Company, 1992.

———. *Our Home Made of Stone: Building in Our Seventies and Nineties.* Camden, Maine: Downeast Books, 1983.

———. *Simple Food for the Good Life.* New York: Delacorte Press, 1980.

———. "Summing Up," *Down East Magazine,* July 1979, 60–63.

———. *Wise Words on the Good Life.* New York: Schocken Books, 1980.

Nearing, Helen, and Scott Nearing. *The Brave New World.* Harborside, Maine: Social Science Institute, 1958.

———. *Building and Using Our Sun-Heated Greenhouse.* Charlotte, Vt.: Garden Way Publishing, 1977.

———. *Continuing the Good Life: Half a Century of Homesteading.* New York: Schocken Books, 1979.

———. *The Good Life: Helen and Scott Nearing's 60 Years of Self-Sufficient Living.* New York: Schocken Books, 1989.

———. *Living the Good Life: Being a Plain Practical Account of a Twenty Year Project in a Self-Subsistent Homestead in Vermont, Together with Remarks on How to Live Sanely and Simply in a Troubled World.* Harborside, Maine: Social Science Institute, 1954.

———. *Living the Good Life: How to Live Sanely and Simply in a Troubled World.* New York: Schocken Books, 1970.

———. *The Maple Sugar Book: With Remarks on Pioneering as a Way of Life in the Twentieth Century.* White River Junction, Vt.: Chelsea Green Publishing Company, 2000. First published in 1950 by John Day and 1970 by Schocken Books.

———. *Socialists Around the World.* New York: Monthly Review Press, 1958.

———. *USA Today: Reporting Extensive Journeys and First-hand Observations Commenting on Their Meaning and Offering Conclusions Regarding Present-Day Trends in the Domestic and Foreign Affairs of the United States.* Harborside, Maine: Social Science Institute, 1955.

Nearing, Robert, with Elena S. Whiteside. "Memory Keeps Me Company." Personal memoir, 1998.

Nearing, Scott. *Black America.* New York: Schocken Books, 1969. First published in 1929 by Vanguard Press.

———. *Civilization and Beyond: Learning from History.* Harborside, Maine: Social Science Institute, 1975.

————. *The Conscience of a Radical.* Harborside, Maine: Social Science Institute, 1965.

————. *Democracy Is Not Enough.* New York: Island Workshop Press, 1945.

————. *Dollar Diplomacy: A Study in American Imperialism.* New York: Monthly Review Press, 1966. First published in 1925 by Viking Press.

————. *Free Born: An Unpublishable Novel.* Freeport, N.Y.: Books for Libraries Press, 1972. First published in 1932.

————. *Freedom: Promise and Menace, A Critique on the Cult of Freedom.* Harborside, Maine: Social Science Institute, 1961.

————. "The Great Madness: A Victory for the American Plutocracy." New York: Rand School, 1917.

————. *The Making of a Radical: A Political Autobiography.* New York: Harper Collins, 1972.

————. *Man's Search for the Good Life.* Harborside, Maine: Social Science Institute, 1954. Reprint 1974.

————. *The Revolution of Our Time.* Washington, D.C.: World Events Committee, 1948.

————. *The Tragedy of Empire.* New York: Island Press, 1945.

————. *The Twilight of Empire: An Economic Interpretation of Imperialist Cycles.* New York: Vanguard Press, 1930.

————. *United World: The Road to International Peace.* New York: Island Press, 1945.

————. *War or Peace?* New York: Island Press, 1946.

Nearing, Scott, and Nellie M. S. Nearing. *Woman and Social Progress: A Discussion of the Biologic, Domestic, Industrial and Social Possibilities of American Women.* New York: Macmillan, 1912.

SECONDARY SOURCES

Abbott, Collamer M. "They Get Away From It All in Pikes Falls," *Vermont Phoenix,* November 12, 1948.

Aley, Jack. "Homesteading Conference: The Enemy Wasn't Even There." *The Maine Times,* May 2, 1975, 36.

Alpern, Sara, Joyce Antler, Elisabeth Israels Perry, and Ingrid Winther Scobie, eds. *The Challenge of Feminist Biography: Writing the Lives of Modern American Women.* Urbana: University of Illinois Press, 1992.

Anderson, Terry H. *The Movement: The Sixties.* New York: Oxford University Press, 1995.

Ascher, Carol, Louise A. De Salvo, and Sara Ruddick, eds. *Between Women: Biogra-*

phers, Novelists, Critics, Teachers and Artists Write about Their Work on Women. Boston: Beacon Press, 1984.

Barthes, Roland. *Camera Lucida: Reflections on Photography.* New York: Hill and Wang, 1981.

Beard, Mary. *The Invention of Jane Harrison.* Cambridge: Harvard University Press, 2000.

Benstock, Shari, ed. *The Private Self: Theory and Practice of Women's Autobiographical Writings.* Chapel Hill: University of North Carolina Press, 1988.

Berry, Wendell. *The Unsettling of America: Culture and Agriculture.* San Francisco: Sierra Club Books, 1977.

"Blazing Your Own Trail to the Good Life." *Speakeasy: A Literary Look At Life* (March/April 2004): 9–45.

Bloom, Alexander, ed. *Long Time Gone: Sixties America Then and Now.* Oxford: Oxford University Press, 2001.

Borsodi, Ralph. *The Flight from the City: An Experiment in Creative Living On the Land.* New York: Harper and Row, 1933.

Braude, Ann. *Radical Spirits: Spiritualism and Women's Rights in Nineteenth-Century America.* Boston: Beacon Press, 1989.

Breines, Wini. *Community and Organization in the New Left 1962–1968: The Great Refusal.* New Brunswick, N.J.: Rutgers University Press, 1989.

Bright, Jean Hay. *Meanwhile, Next Door to the Good Life.* Dixmont, Maine: Bright-Berry Press, 2003.

Bromfield, Louis. *Pleasant Valley.* New York: Harper and Brothers, 1945.

Brown, James P. "Scott Nearing: 100 Years of 'The Good Life,'" *Down East Magazine* 30 (August 1983): 171.

Buck, Pearl S. *The Good Earth.* New York: John Day Company, 1931.

Buell, Lawrence. *The Environmental Imagination: Thoreau, Nature Writing, and the Formation of American Culture.* Cambridge, Mass.: Belknap Press of Harvard University Press, 1995.

Burroughs, John. *Signs and Seasons.* Boston: Houghton Mifflin and Company, 1886.

Butsch, Richard. "Introduction: Leisure and Hegemony in America," in *For Fun and Profit: The Transformation of Leisure into Consumption,* ed. Richard Butsch. Philadelphia: Temple University Press, 1990.

Campbell, Bruce F. *Ancient Wisdom Revived: A History of the Theosophical Movement.* Berkeley: University of California Press, 1980.

Canfield, Clarke. "Nearing Spirit Lives On," *Bangor Daily News,* September 22–23, Section G, 1–2.

Carpenter, Murray. "The Riddle of Self-Sufficiency," *The Maine Times*, August 12, 1999, 4–9.

Carr, David. "Narrative and the Real World: An Argument for Continuity," in *Memory, Identity, Community: The Idea of Narrative in the Human Sciences*, ed. Lewis P. Hinchman and Sandra K. Hinchman. Albany: SUNY Press, 1997.

———. *Time, Narrative, History*. Bloomington: Indiana University Press, 1986.

"Carrying on the Good Life," *Down East Magazine* 32, no. 42 (November 1985).

Carson, Rachel. *Silent Spring*. Boston: Houghton Mifflin, 1962.

Case, John, and Rosemary C. R. Taylor, eds. *Co-Ops, Communes, and Collectives: Experiments in Social Change in the 1960s and 1970s*. New York: Pantheon Books, 1979.

Chiras, Daniel D. *Beyond the Fray: Reshaping America's Environmental Response*. Boulder: Johnson Books, 1990.

Clark, Claudia. *Radium Girls: Women and Industrial Health Reform, 1910–1935*. Chapel Hill: University of North Carolina Press, 1997.

Clark, Jeff. "Summing Up the Good Life," *Down East Magazine*, April 1992, 22–30.

Clarke, John. "Pessimism versus Populism: The Problematic Politics of Popular Culture," in *For Fun and Profit: The Transformation of Leisure into Consumption*, ed. Richard Butsch. Philadelphia: Temple University Press, 1990.

Coleman, Eliot. *The New Organic Grower: A Master's Manual of Tools and Techniques for the Home and Market Gardener*, rev. and exp. edition. White River Junction, Vt.: Chelsea Green Publishing Company, 1999.

Commoner, Barry. *The Poverty of Power: Energy and the Economic Crisis*. London: Jonathon Cape, 1976.

Conzen, Michael P., ed. *The Making of the American Landscape*. Boston: Unwin Hyman, 1990.

Copperthwaite, William S. *A Handmade Life: In Search of Simplicity*. White River Junction, Vt.: Chelsea Green Publishing Company, 2002.

Cullen, Jim. *The Art of Democracy: A Concise History of Popular Culture in the United States*. New York: Monthly Review Press, 1996.

Curti, Lidia. *Female Stories, Female Bodies: Narrative, Identity and Representation*. New York: New York University Press, 1998.

Day, Dorothy. *House of Hospitality*. New York: Sheed and Ward, 1939.

Denning, Michael. *The Cultural Front: The Laboring of American Culture in the Twentieth Century*. London: Verson, 1996.

Diamond, Irene, and Gloria Feman Orenstein, eds. *Reweaving the World: The Emergence of Ecofeminism*. San Francisco: Sierra Club Books, 1990.

Elgine, Duane. *Voluntary Simplicity: Toward a Way of Life That Is Outwardly Simple, Inwardly Rich.* New York: William Morrow and Company, 1981.

Ellsberg, Robert, ed. *By Little and By Little, The Selected Writings of Dorothy Day.* New York: Alfred A. Knopf, 1983.

Encyclopedia of the American Left. Ed. Mari Jo Buhle, Paul Buhle, and Dan Georgakas. New York: Garland Publishing, 1990.

Engelhardt, Tom. *The End of Victory Culture: Cold War America and the Disillusioning of a Generation.* New York: Basic Books, 1995.

Evans, Sara M. *Born for Liberty: A History of Women in America.* New York: The Free Press, 1989.

Ewen, Stuart. *Captains of Consciousness: Advertising and The Social Roots of the Consumer Culture.* New York: McGraw-Hill, 1976.

Fox, Stephen. *John Muir and His Legacy: The American Conservation Movement.* Boston: Little, Brown and Company, 1981.

Freeman, Mark, and Jens Brockmeier. "Narrative Integrity: Autobiographical Identity and the Meaning of the 'Good Life,'" in *Narrative and Identity: Studies in Autobiography, Self and Culture,* ed. Jens Brockmeier and Donald Carbaugh. New York: John Benjamins Publishing Company, 2001.

Gaard, Greta, ed. *Ecofeminism: Women, Animals, Nature.* Philadelphia: Temple University Press, 1993.

Garrett, Richard. "What the Nearings Have Meant to Us," *Country Journal,* 1979.

Gitlin, Todd. *The Sixties: Years of Hope, Days of Rage.* New York: Bantam Books, 1987.

Glassie, Henry. "Folkloristic Study of the American Artifact: Objects and Objectives," in *Folkloristic Study of the American Artifact,* ed. Dorson Carpenter and Gale Carpenter. Bloomington: Indiana University Press, 1983.

Goodyear, Dana. "The Simple Life, Inc.: How MaryJane Butters Reinvented the Farm Girl." *New Yorker Magazine,* October 11, 2004.

Gould, Rebecca Kneale. *At Home in Nature: Modern Homesteading and Spiritual Practice in America.* Berkeley: University of California Press, 2005.

———. "At Home in Nature: The Religious and Cultural Work of Homesteading in Twentieth-Century America." PhD diss., Harvard University, 1997.

Graham, David L. "The Ruggedest, Individualest Man in Maine," *The Maine Times,* June 6, 1969, 8–9.

Greenberg, Paul. "The Really Simple Life: How Two Big-City Radicals Helped Take America Back to the Land." *Boston Sunday Globe,* September 5, 2004.

Gregg, Richard. *The Power of Non-Violence.* Philadelphia: J.B. Lippincott Company, 1966. First published in 1934 by J.B. Lippincott.

Gregg, Richard, and Helen Louise Porter Phillbrick. *Companion Plants and How to Use Them*. Old Greenwich, Conn.: Devon-Adair Company, 1966.

Groth, Paul, and Todd Bressi, eds. *Understanding Ordinary Landscapes*. New Haven: Yale University Press, 1997.

Grumbach, Doris. *Life in a Day*. Boston: Beacon Press, 1996.

Gunst, Kathy. "Ripping Apart the American Style of Eating," *The Maine Times*, May 15, 1987, 27.

Hambidge, Gove. *Enchanted Acre*. New York: McGraw-Hill, 1935.

Hart, John Fraser. *The Look of the Land*. Englewood Cliffs, N.J.: Prentice-Hall, 1975.

Hayden, Dolores. *The Grand Domestic Revolution: A History of Feminist Designs for American Homes, Neighborhoods and Cities*. Cambridge, Mass.: MIT Press, 1981.

Hays, Samuel P. *Beauty, Health and Permanence: Environmental Politics in the United States, 1955–1985*. Cambridge: Cambridge University Press, 1987.

Hedgepath, William. *The Alternative: Communal Life in New America*. London: Collier MacMillan, 1970.

Heilbrun, Carolyn G. *Writing a Woman's Life*. New York: Ballantine Books, 1988.

Helen Nearing: Conscious Living/Conscious Dying. Produced by Polly Bennell and Andrea Sarris. VHS. Bullfrog Films, 2000.

Hinchman, Lewis P., and Sandra K. Hinchman, eds. *Memory, Identity, Community: The Idea of Narrative in the Human Sciences*. Albany: SUNY Press, 1997.

Houriet, Robert. *Getting Back Together*. New York: Coward, McCann, and Geoghegan, 1971.

Isaac, Rhys. *The Transformation of Virginia, 1740–1790*. New York: W.W. Norton, 1988.

Jackson, Shannon. *Lines of Activity: Performance, Historiography, Hull-House Domesticity*. Ann Arbor: University of Michigan Press, 2000.

Jackson, W. Charles. "Quest for the Good Life: Helen and Scott Nearing and the Ecological Imperative." Masters Thesis, University of Maine, 1993.

Jacob, Jeffrey. *New Pioneers: The Back-to-the-Land Movement and the Search for a Sustainable Future*. University Park: Pennsylvania State University, 1997.

Johnson, Linnea. "Nearing Enough," *Mother Earth News* (October/November 2003): 22–27.

Joly, Greg. "Epilogue," *The Maple Sugar Book: With Remarks on Pioneering as a Way of Life in the Twentieth Century*. White River Junction, Vt.: Chelsea Green Publishing, 2000.

———. *A Love Greater than 70 Bushels of Baked Potatoes: Helen and Scott Nearing in Vermont*. A One-Act Play. Performed at the Stratton Mountain Summer Symposium, June 23, 2002.

Joseph, Stanley. *Maine Farm: A Year of Country Life.* New York: Random House, 1991.

Judd, Richard, and Christopher S. Beach. *Natural States: The Environmental Imagination in Maine, Oregon, and the Nation.* Washington, D.C.: Resources for the Future, 2003.

Juliette of the Herbs. Mabinogion Films production, directed by Tish Streeten. VHS. Wellspring Media, 1998.

Kammen, Michael G. *American Culture, American Tastes: Social Change and the 20th Century.* New York: Alfred A. Knopf, 1999.

Kelly, William C. "Rodale Press and Organic Gardening," *HortTechnology,* April/June 1992.

Kirshenblatt-Gimblett, Barbara. *Destination Culture: Tourism, Museums and Heritage.* Berkeley: University of California Press, 1998.

Kyle, Richard. *The New Age Movement in American Culture.* Lanham, Md.: University Press of America, 1995.

LaConte, Ellen. *On Light Alone: A Guru Meditation on the Good Death of Helen Nearing.* Stockton Springs, Maine: Looseleaf Press, 1996.

Lear, Linda J. "Rachel Carson's *Silent Spring,*" *Environmental History Review,* Summer 1993.

Lears, T. J. Jackson. "The Concept of Cultural Hegemony: Problems and Possibilities," *American Historical Review* (June 1985): 567–93.

———. *Fables of Abundance: A Cultural History of Advertising in America.* New York: Basic Books, 1994.

———. *No Place of Grace: Antimodernism and the Transformation of American Culture 1880–1920.* New York: Pantheon Books, 1981.

Leopold, Aldo. *A Sand County Almanac: With Other Essays on Conservation from Round River.* New York: Ballantine Books, 1970. First published by Oxford University Press in 1949.

Lepore, Jill. "Historians Who Love Too Much: Reflections on Microhistory and Biography," *Journal of American History* (June 2001): 129–44.

Levine, Lawrence W. "The Folklore of Industrial Society: Popular Culture and Its Audiences," *American Historical Review* (December 1992): 1369–99. Responses by Robin Kelley, Natalie Zemon Davis, Jackson Lears, and Levine, 1400–30.

Lewis, Peirce F. "Axioms for Reading Landscape," in *The Interpretation of Ordinary Landscapes: Geographical Essays,* ed. D. W. Meinig. New York: Oxford University Press, 1979.

Living the Good Life. VHS. Bullfrog Films, 1977.

Loomis, Mildred. "Vermont Homesteading for Living the Good Life," *The Interpreter,* February 1955.

Lutts, Ralph H. "Chemical Fallout: Rachel Carson's *Silent Spring*, Radioactive Fallout, 137
and the Environmental Movement," *Environmental Review* (Fall 1985): 211–25.

Lutyens, Mary. *Krishnamurti: His Life and Death.* New York: St. Martin's Press, 1990.

———. *Krishnamurti: The Years of Awakening.* New York: Farrar, Straus, and Giroux, 1975.

May, Elaine Tyler. "Cold War—Warm Hearth: Politics and Family," in *The Rise and Fall of the New Deal Order, 1930–1980,* ed. Steve Fraser and Gary Gerstle. Princeton: Princeton University Press, 1989.

McAdams, Dan P. *The Stories We Live By: Personal Myths and the Making of the Self.* New York: William Morrow and Company, 1993.

McCarthy, Colman. "Two Escapees from Get-Aheadism," *Washington Post,* September 23, 1995.

McQuiston, John. "Helen K. Nearing, Maine Author, Dies at 91," *New York Times,* September 19, 1994, B8.

Melville, Keith. *Communes in the Counter Culture: Origins, Theories, Styles of Life.* New York: William Morrow and Company, 1972.

Merchant, Carolyn. *The Death of Nature: Women, Ecology, and the Scientific Revolution.* San Francisco: Harper and Row, 1980.

———. "Ecofeminism and Feminist Theory," in *Reweaving the World: The Emergence of Ecofeminism,* ed. Irene Diamond and Gloria Feman Orenstein. San Francisco: Sierra Club Books, 1990.

———. *Radical Ecology: The Search for a Livable World.* New York: Routledge, 1992.

Mishler, Elliot G. "Models of Narrative Analysis: A Typology," *Journal of Narrative and Life History* (1995): 87–123.

The Nearings at Forest Farm: The Vermont Years. Produced by Sandy Mackinnon, C. J. King, and Greg Joly. VHS. Stratton Foundation's 2002 Summer Symposium.

Opie, John. *Nature's Nation: An Environmental History of the United States.* Fort Worth: Harcourt Brace College Publishers, 1998.

Personal Narratives Group, eds. *Interpreting Women's Lives: Feminist Theory and Personal Narratives.* Bloomington: Indiana University Press, 1989.

Petulla, Joseph M. *Environmental Protection in the United States.* San Francisco: San Francisco Study Center, 1987.

Plant, Judith, ed. *Healing the Wounds: The Promise of Ecofeminism.* Philadelphia: New Society Publishers, 1989.

Pollock, Griselda. *Vision and Difference: Femininity, Feminism and Histories of Art.* New York: Routledge, 1988.

Polson, Sheila. "Back-to-the-Land Pioneers Preserve Their Legacy," *Christian Science Monitor,* January 4, 1996.

"The Plowboy Interview." *Mother Earth News,* 1971, 6–12.

Raver, Anne. "Helen K. Nearing: Greener, Saner, Simpler," *New York Times,* December 31, 1995, 35.

Review of *Living the Good Life, Down East Magazine* 17 (November 1970): 88.

Review of *Living the Good Life. The Maine Times,* April 20, 1973.

Richmond, Al. *A Long View from the Left.* Boston: Beacon Press, 1973.

Rose, Vivien Ellen. "Homesteading as Social Protest: Gender and Continuity in the Back-to-the-Land Movement in the United States 1890–1980 (Helen Nearing, Scott Nearing, Ralph Borsodi, Myrtle May Borsodi, Mildred Loomis)." PhD diss. at SUNY Binghamton, 1997.

Rosenberg, Norman L., and Emily S. Rosenberg. *In Our Times: America since World War II,* fifth edition. Englewood Cliffs, N.J.: Prentice Hall, 1995.

Rothman, Hal K. *The Greening of a Nation? Environmentalism in the United States Since 1945.* New York: Harcourt Brace Publishers, 1998.

Ryden, Kent. *Mapping the Invisible Landscape: Folklore, Writing and the Sense of Place.* Iowa City: University of Iowa Press, 1993.

St. George, Robert. *Material Life in America 1600–1860.* Boston: Northeastern University Press, 1988.

Sale, Kirkpatrick. *The Green Revolution: The American Environmental Movement, 1962–1992.* New York: Hill and Wang, 1993.

———. *Human Scale.* New York: Coward, McCann and Geoghegan, 1980.

Saltmarsh, John A. *Scott Nearing: An Intellectual Biography.* Philadelphia: Temple University Press, 1991.

Scarce, Ric. *Eco-Warriors: Understanding the Radical Environmental Movement.* Chicago: Noble Press, 1990.

Scarry, Elaine. *The Body in Pain: The Making and Unmaking of the World.* New York: Oxford University Press, 1985.

Schumacher, E. F. *Good Work.* New York: Coward, McCann and Geoghegan, 1980.

Shannon, David. *The Decline of American Communism.* Chatham, N.J.: Chatham Bookseller, 1971.

Shi, David E. *In Search of the Simple Life: American Voices, Past and Present.* Salt Lake City: Peregrine Smith Books, 1986.

———. *The Simple Life: Plain Living and High Thinking in American Culture.* New York: Oxford University Press, 1985.

Shiva, Vandana, ed. *Biodiversity Conservation: Whose Resource? Whose Knowledge?* New Delhi: Indian National Trust for Art and Cultural Heritage, 1994.

———. *Staying Alive: Women, Ecology and Development.* London: Zed Books, 1988.

Sloss, Radha Rajagopal. *Lives in the Shadow with J. Krishnamurti.* Reading, Mass.: Addison-Wesley Publishing Company, 1991.

Smith, Sidonie, and Julia Watson, eds. *Getting a Life: Everyday Uses of Autobiography.* Minneapolis: University of Minnesota Press, 1999.

———. *Women, Autobiography, Theory: A Reader.* Madison: University of Wisconsin Press, 1998.

Smith-Rosenberg, Carroll. *Disorderly Conduct: Visions of Gender in Victorian America.* New York: Alfred A. Knopf, 1985.

Starobin, Joseph. *American Communism in Crisis 1943–1957.* Cambridge, Mass.: Harvard University Press, 1972.

Terkel, Studs. *American Dreams: Lost and Found.* New York: Pantheon Books, 1980.

Thoreau, Henry David. *Walden.* New York: Bramhall House, 1951.

Toynbee, Arnold. *Civilization on Trial.* New York: Oxford University Press, 1948.

Trachtenberg, Alan. *Reading American Photographs: Images as History Mathew Brady to Walker Evans.* New York: Hill and Wang, 1989.

Tucker, Anne Wilkes, Claire Cass, and Stephen Daiter, eds. *This Was the Photo League: Compassion and the Camera from the Depression to the Cold War.* Chicago: Stephen Daiter Gallery, 2001.

Tumber, Catherine. *American Feminism and the Birth of New Age Spirituality: Searching for the Higher Self, 1875–1915.* Lanham, Md.: Rowman and Littlefield Publishers, 2002.

Twigg, Reginald. "The Performative Dimension of Surveillance: Jacob Riis' *How the Other Half Lives,*" *Text and Performance Quarterly* (October 1992): 305–28.

Upton, Dell. "Outside the Academy: A Century of Vernacular Architecture Studies, 1890–1990," in *The Architectural Historian in America,* ed. Elizabeth Blair Mac-Dougall. Washington, D.C.: National Gallery of Art, 1990.

Wald, Alan M. *The New York Intellectuals.* Chapel Hill: University of North Carolina Press, 1987.

Walsh, Pamela M. "Helen Nearing, Popular Author Who Praised Simple Living, at 91," *Boston Globe,* September 18, 1995.

Warren, Frank A., III. *Liberals and Communism.* Bloomington: Indiana University Press, 1966.

Washington, Peter. *Madame Blavatsky's Baboon: A History of the Mystics, Mediums and Misfits Who Brought Spiritualism to America.* New York: Schocken Books, 1995.

White, Hayden V. *The Content of the Form: Narrative Discourse and Historical Representation.* Baltimore: Johns Hopkins University Press, 1987.

140   Whitfield, Stephen J. *Scott Nearing: Apostle of American Radicalism.* New York: Columbia University Press, 1974.

Wu, Yuan-Li. *Tiananmen to Tiananmen: China Under Communism 1947–1996, After Delusion and Disillusionment A Nation at a Crossroads.* Occasional Papers/Reprints Series in Contemporary Asian Studies, Number 1. College Park: School of Law, University of Maryland, 1997.

Zuckerman, Michael. *Almost Chosen People: Oblique Biographies in the American Grain.* Berkeley: University of California Press, 1993.

)

(Continued from page iv)

Gale Disney, interview with Margaret Killinger, tape recording, March 4, 1999.

Richard Garret, interview with Margaret Killinger, May 17, 2006.

Jeanne Gaudette, interview with Margaret Killinger, tape recording, March 3, 1999.

Jerry Goldman, interview with Margaret Killinger, tape recording, June 14, 2002.

Rebecca and Eugene Lepkoff, interview with Margaret Killinger, tape recording, December 29, 2001.

Cornelia Tuttle, telephone interview with Margaret Killinger, March 14, 2004.

Written material

Jane M. Hamel, letter to Helen Nearing dated May 19, 1993.

Kathryn Kester, letter to Helen Nearing dated October 22, 1992.

Yukio Okamura, letter to Helen Nearing dated September 9, 1993.

Les Oke, letter to Helen Nearing dated January 28, 1993.

Pete Seeger, lyrics from "Maple Syrup Time," © 1977 by Sanga Music, Inc. (Full credit appears on page 67.)

Photograph credits appear with the respective captions.

Index 🌿